PENGUIN BOOKS

VIRTUAL WORLDS

Benjamin Woolley introduced the idea of virtual reality to Britain with his coverage of the subject for the *Listener* magazine in 1988 and in an item which he presented for BBC2's arts and culture programme *The Late Show* in 1989. As well as being a correspondent for *The Late Show* and writing 'Signs of Life', an edition of *Horizon* on the use of computers to study and simulate life forms, he has also contributed to a number of national newspapers and magazines on technology, design and the arts.

Benjamin Woolley won the 1991 BP Arts Journalism television award. *Virtual Worlds* was shortlisted for the 1993 Science Book Prize.

BENJAMIN WOOLLEY

VIRTUAL WORLDS

A JOURNEY IN HYPE AND HYPERREALITY

PENGUIN BOOKS

PENGUIN BOOKS

Published by the Penguin Group
Penguin Books Ltd, 27 Wrights Lane, London W8 5TZ, England
Penguin Books USA Inc., 375 Hudson Street, New York, New York 10014, USA
Penguin Books Australia Ltd, Ringwood, Victoria, Australia
Penguin Books Canada Ltd, 10 Alcorn Avenue, Toronto, Ontario, Canada M4V 3B2
Penguin Books (NZ) Ltd, 182–190 Wairau Road, Auckland 10, New Zealand

Penguin Books Ltd, Registered Offices: Harmondsworth, Middlesex, England

First published by Blackwell 1992
Published in Penguin Books 1993
1 3 5 7 9 10 8 6 4 2

Printed in England by Clays Ltd, St Ives plc

When you're interacting with a computer, you are not conversing with another person. You are exploring another world.

John Walker

CONTENTS

ACKNOWLEDGEMENTS

Like any other, this book is, in a sense, a 'hypertext', a text created out of other texts, out of the links that join the great body of work concerning the areas it touches. Therefore it is to those hundreds of writers, technologists and academics who created that body of work that I owe the greatest debt of thanks. More specific acknowledgement must go to *The Listener* and *The Late Show*, both of which succoured the ideas that follow with their adventurous commitment to areas that other current affairs magazines and arts programmes dared not cover. I would also like to thank John Wyver for allowing me to benefit from his unerring knack for spotting interesting new ideas and work, David Albert for his views on some of the tricky cosmological issues raised in chapter 11, and staff at Blackwell Publishers, particularly Stephan Chambers, for helping me through the unfamiliar territory beyond the journalist's 1,000-word event horizon.

This book is dedicated to *The Listener*, which closed on 3 January 1991 after 62 years of publication.

INTRODUCTION

In our days everything seems pregnant with its contrary.
                                        Karl Marx, 1856.[1]

On 7 May, 1987, the US multinational Proctor & Gamble submitted Olestra, a new food substitute, to the American Food and Drug Administration (FDA) for approval. Olestra was, in the judgement of the world's media as well as Proctor & Gamble's publicity, potentially one of the most significant nutritional breakthroughs of its time. It promised what every dieter desired, the realization of an impossible dream: fat-free fat.

For a dietary aid, Olestra is made out of unpromising ingredients, sugar and fat. However, when they are chemically bonded together in the correct way, they form a new and very strange substance: sucrose polyester. It may sound better suited to manufacturing shirts than food, but sucrose polyester has interesting nutritional qualities. It retains the culinary and textural qualities of the fat, but in a form that the body is unable to digest. Result: a fat that passes straight through the body. In tests, obese subjects who ate a diet that used Olestra lost weight even if they were allowed to supplement the diet with conventional fatty snacks.[2]

Fat-free fat. It is a concept pregnant with its own contrary. When Marx used the phrase in 1856, he was commenting on an era that was having to come to terms with the violent impact of industrialization on the order of nature. This book is about the impact of an even more disorientating era: artificialization, of making dreams come true, of 'imagineering' as the Disney

Corporation calls it. What are the extent and limits of the artificial? Is there, can there be, any contact with reality when it is possible to make fat that is not fat, when the fake becomes indistinguishable from – even more authentic than – the original, when computers can create synthetic worlds that are more realistic than the real world, when technology scorns nature?

Evidence of artificialization seemed to be abundant at the time of writing this book. Wherever one looked, artificiality was triumphing over reality. In the closing months of 1991, as the Soviet Union broke apart, the fate of the embalmed body of Lenin, which had been lying in state since 1924, was in doubt. Some had questioned whether, indeed, it was his 'real' body at all, Lenin's unblemished facial complexion suggesting that it might be a waxwork. But who cares? Most British politicians would happily become a waxwork if it meant a position at Madame Tussaud's, the most popular tourist site in London. In an artificial world, there is no need for Lenin's material substance any more than there is a need for the great materialist state he founded. It is, perhaps, a fitting symbol of the triumph of the land of dreams over the empire of iron, the 'end of history', that Lenin should be left to decay while the cryogenically preserved Walt Disney waits in suspended animation (an appropriate state for a cartoonist) to return to the world that is now his.

As the worms awaited Lenin, a book entitled *Fly Fishing* by J. R. Hartley entered the British bestseller list. Unremarkable though it may have seemed, this was no ordinary book. British Telecom had run a series of television commercials featuring an elderly man locating a copy of the out-of-print *Fly Fishing* by J. R. Hartley using the Yellow Pages. The man turns out at the end of the commercial to be J. R. Hartley. So moved was the audience by this story that bookshops and even the British Library were reportedly overwhelmed with requests for it – even though, of course, no such book and no such author had ever existed.[3] So the publisher Random Century decided to create one. It commissioned the writer Michael Russell to ghost write the book, and hired the actor Norman Lumsden, who had played Hartley in the original advertisement, to pose as the author. The result was a fiction turned into fact – artificial reality.

Few of us think about 'reality' much – those of us who intend to write books about it have to keep reminding ourselves that we are rare exceptions. It is, perhaps, the conceptual equivalent of unconscious motor functions such as breathing. It is vital to life – without it, we would be unable to distinguish the real from the imaginary, the true from the false, the natural from the artificial. But we do not have to think about it to use it – indeed, as soon as we do start thinking about it, it becomes extremely difficult to continue using it. For this reason, perhaps, some may regard it as a peculiar subject for any sort of analysis: it is a given, a fact of life, and best left hidden behind the curtain of unconsciousness.

To extend (but I hope not exhaust) the breathing metaphor, the problem is that just as a polluted environment can make us short of breath, so an increasingly complex, artificial environment can diminish our sense of reality. And, as that sense diminishes, so innumerable troubling side-effects start to creep in. Soon after Olestra's announcement, an American pressure group, the Center for Science in the Public Interest, quoted the results of tests using rats which showed that nearly a half of the rats fed on Olestra died during the period of the experiment, compared with under a third of the rats fed a normal diet.[4] 'Aha!' said the critics, 'There you have it: you think you can have fat for free, but there is really a price, the price that nature usually exacts, cancer.'

The prejudice that favours the products of nature over our own is, perhaps, understandable. Nature's approvals process is slower even than the FDA's, working at the pace of evolutionary time to separate dangerous substances from those to whom they are a danger. We are part of a natural order poised in a state of delicate equilibrium, safe as long as we keep to our position within it. This 'natural order' is a very basic, important structure. It is independent of us, uncontaminated by us. The problem with synthetic substances like Olestra is that they are not a part of it – worse, that they seem to ignore it, even violate it. Fat-free fat, like alcohol-free alchoholic drinks, sugar-free sweets and caffeine-free coffee, flouts the reality principle, the principle that you cannot have something for nothing, that everything has its price, that nothing in life is free, that there is no such thing as a free lunch, that there can be no gain without pain. Science and technology have arrogantly ignored this principle, and the result is a world filled with disease and pollution.

But must it be thus? Surely there *could* be gain without pain – or, more appropriately given that Olestra is a dietary aid, less without distress, crapulence without corpulence? As the *US News & World Report* put it in a rather breathless report about food substitutes entitled 'Have your cake and eat it, too', 'the foodstuffs of which dieters dream are fast becoming realities'.[5] And why not? A few cancerous rats prove nothing: there is no law of science (though perhaps one of logic) that says a fat-free fat is impossible.

It is this sort of debate, and what for most of us is a genuine feeling of uncertainty as to exactly what we can believe that makes any secure, unexamined notion of reality increasingly troublesome. How is it possible to hold a clear view of the distinction between reality and fantasy when the unreal is continually being realized?

There has never been a totally secure view of reality, certainly not in the industrial era of history. People say that the world is not as real than it used to be. Well, to adapt what an editor of *Punch* said in response to a tiresome criticism of his magazine: the world was *never* as real as it used to be. Indeed, it is industry, the power to manufacture what previously had to be taken from nature, that has made the world progressively more artificial and less real, that provided the wealth and energy to change the natural landscape, even to replace it with one of our own making. But equally the industrial era has been about the discovery of reality. The ability to manipulate nature, to turn its operation to our own ends, shows how successful science has been in discovering how it works, and technology in exploiting that discovery. The industrial experience, in other words, seems to have both destroyed reality and reinforced it.

As we enter the so-called post-industrial era, the crisis continues. It is a favourite theme of nearly all commentaries about the times we live in. There is a 'legitimation crisis',[6] a 'crisis of representation',[7] and one great big 'crisis of modernity'. As Umberto Eco observed: 'Crisis sells well. During the last few decades we have witnessed the sale (on newsstands, in book-shops, by subscription, door-to-door) of the crisis of religion, of Marxism, of representation, the sign, philosophy, ethics, Freudian-ism, presence, the subject. . . . Whence the well-known quip: "God is dead, Marxism is undergoing crisis, and I don't feel so hot myself."'[8] The question such crises pose is whether this means that attitudes to reality have been undermined by the experience of modernity, or whether reality itself, something firm and objective, something underpinning the uncertain world of appearance, has been shown to be an illusion. Is the lesson we should have learnt from the last century that *there is no reality*? When the newspapers and the food manufacturers tell us that dreams are fast becoming realities, does that really mean that reality is fast becoming a dream?

For centuries, the issue of what does and does not count as real has been a matter of philosophy. There have been two main questions, what could be called the ontological and epistemological questions. The ontological question is about being: what is real? Is there a reality behind appearance? The epistemological question is about knowing: what is truth? Is knowledge the product of reason or of experience? For most people, and certainly for most scientists, neither question is particularly relevant, because there are working systems that are used to tell reality from appearance, truth from falsity. When a court assesses the truth of a witness's evidence, epistemology does not come into it. The judge does not ask the members of the jury whether or not they are rationalists or empiricists. Similarly, the result of an experiment does not rest on whether the scientist performing the experiment is an idealist or materialist.

At Britain's first virtual reality conference, held in the summer of 1991, the chairman, Tony Feldman, tried to emphasize that the agenda he had drawn up was concerned with hard-headed, pragmatic business issues, but conceded that 'the metaphysics are inescapable'.[9] Technology, he and the subsequent speakers observed, could manipulate reality to the point of being able to create it. Artificialization is no longer just a matter of cultural observation or intellectual angst, it had become, well, real. It is for this reason that reality is no longer secure, no longer something we can simply assume to be there.

Most people probably now have some idea of what virtual reality is. It is the technology used to provide a more intimate 'interface' between humans and computer imagery. It is about simulating the full ensemble of sense data that make up 'real' experience. Ideally, the user wears a device that subsitutes the sense data coming from the natural world with that produced by a computer. Computer screens are placed before the eyes, 'effectors' cover the body, providing the sights of this artificial world, and the feelings that result from touching it. Furthermore, tracking devices attached to the body monitor its movements, so, as the user moves, so what he or she sees and feels is altered accordingly. This book does not provide a technical description or assessment of this technology. I am more concerned with the two issues that underlie its emergence as one of the 'Big Ideas' of the 1990s, issues that remain neglected by the computer industry, but which I hope to show are essential to making sense of the developments that have led to this rude instrusion of metaphysics into ordinary life.

The first issue is simulation. Computers are unique in that they are all, in a sense, simulations of some ideal computer, a 'universal machine'. Everything a computer does can be seen as a simulation, except that many of the things it simulates do not exist beyond the simulation. What, then, is simulation? Is it just another form of imitation or representation, fiction for the computer age? Can anything be simulated – even reality and human intelligence? These questions raise important mathematical and scientific questions, and in attempting to answer them I hope to show what is 'special' about the computer, why it is not just a glorified calculator, and why those who have developed it have attributed to it is such extraordinary creative powers.

The second issue concerns artificial reality. It is a strange, provocative notion. It is hard to tell on first acquaintance whether it is meaningful or meaningless. It could be a new paradigm, it could be pretentious. It could be an oxymoron, a figure of speech that uses what sounds like a contradiction to suggest a much deeper truth. It could just be a contradiction. It has become a general-purpose metaphor for both the present and the future, one that is easy to pick up but impossible to put down. Its attractions are obvious. It is a term full of novelty and puzzle, provocatively coined to intensify a deep-seated insecurity as well as capture a sense of technological adventure.

The term's origin is generally attributed to Myron Krueger, an American 'computer artist and educator' (his own description), who used it as the title of a book written in the 1970s but not published until 1983. The book's subject was what he called 'responsive environments', art installations in which lighting and sound would change according to the movement of people walking around them. He did not, however, use the term 'artificial reality' as a technological label. He had a more ambitious use for it: 'The world described in Genesis, created by mysterious cosmic forces, was a volatile and dangerous place. It moulded human life through incomprehensible caprice. Natural beneficence tempered by natural disaster defined reality. For centuries, the goal of human effort was to tap Nature's terrible power. Our success has been so complete, that a new world has emerged. Created by human ingenuity, it is an artificial reality.[10]

The champions of virtual reality and computer simulation – who are not just pioneering visionaries but powerful commercial interests – are wanting to make artificial reality real. To them, it promises not just a world where you can eat fat without getting fat, not just a metaphor, but the actual creation of any world you could ever want or imagine – fantastical, fabulous, terrifying, infinite, enclosed, utopian, Stygian. I want to look at the validity of such promises, and to discover their influence over the whole notion of reality. Artificial reality has acquired the role of a sort of Barium meal ingested by the body of society and culture, its spreading glow revealing, under the X-ray of critical examination, the growths and malfunctions of the internal organs.

Just like a Barium meal, some people find any discussion about artificial or virtual reality hard to swallow. Upon introducing the idea to innocent members of the lay public, I discovered that, far from showing mild interest, or no interest, they would look at me as though I had announced myself to be the risen Messiah. Life, they seemed to say, is complicated enough without making it more complicated with such outlandish ideas. Such an attitude has a proud and long tradition in Anglo-Saxon culture. You could say it was formulated as long ago as the fourteenth century, when it came to be known as 'Occam's razor', after the English philosopher William of Occam. Occam was a ferociously ascetic Franciscan, so ascetic he actually led a revolt in favour of poverty, when Pope John XXII threatened to end it as a monastical principle. He was an equally ferocious intellectual ascetic, demanding that in philosophical theory, 'entities are not to be multiplied beyond necessity',[11] a principle more economically expressed in the phrase 'cut the crap'. Judged in such terms, most contemporary theory, especially that coming from the fields of information technology and cultural criticism, seems in need of a good shave.

Some contemporary theorists positively encourage the multiplication of entities. One of the most enthusiastic is the influential French sociologist Roland Barthes, who died in 1980. His most important work was *Mythologies*, a book that unblushingly used the most sophisticated analytical techniques to examine the most commonplace objects and activities: wrestling, margarine, photos of Greta Garbo, polystyrene. It was Barthes's explicit aim to break the illusion of 'naturalness' that is used by mass media to dress up reality. He wanted to show that our notion of the natural and the real is really a highly political construction, a product of history: 'In short, in the account given of our contemporary circumstances . . . I wanted to track down, in the decorative display of *what-goes-without-saying*, the ideological abuse which, in my view, is hidden there.'[12] He set out to be reality's party pooper, to show that, in his language, it is the product of myth.

Neologisms are, Barthes claims, essential precisely because of this myth. What we take to be fixed and certain is, in fact, constantly changing, and so the language used to analyse it must change too. The words that appear in dictionaries, words that are presented as having meaning independent of history, are no use. We need new ones: 'neologism is therefore inevitable. China is one thing, the idea which a French petit-bourgeois could have of it not so long ago is another: for this peculiar mixture of bells, rickshaws and opium-dens, no other word is possible but *Sininess*. Unlovely? One should at least get some consolation from the fact that conceptual neologisms are never arbitrary: they are built according to a highly sensible proportional rule'.[13] Artificial reality. Unlovely, or built according to a highly sensible proportional rule? Just another example of intellectual stubble and a candidate for Occam's cut-throat, or a perfectly legitimate example of a new term for something that has not been previously recognized or expressed?

My belief is that artificial reality does reveal a great deal about the 'myth' of reality – about the way that the idea of reality is used and understood, at least within the Western culture that gave birth to it. If nothing else, it reveals that much of what we take to be reality *is* myth, just as Olestra reveals that the idea of fat is a myth. It reveals that the things we assume to be independent of us are actually constructed by us. It reveals that being 'real', like being 'natural', is not simply a value-free, unproblematic, apolitical, objective state – though part of its mythology is to make itself appear to be so. It reveals that, like 'new' and, indeed, 'natural', 'real' has been abducted by business as a marketing term.

Artificial reality, then, expresses the ambiguity of current attitudes to reality. But that ambiguity is not, as most commentators on the subject have taken it to be, evidence that there is no reality. Just because there is a reality myth does not mean that reality is a myth. The absurdity of such a position is revealed in attempts by some antirealists to argue that they cannot assert reality to be a myth because that would be to assert that what is real is that reality is a myth, which cannot be asserted as there is no reality (because it is a myth).

I want to show that such denials of reality are mistaken, that there is a reality, and that the virtual form of it, far from releasing us from it, can help us recover it. I also believe we need it. Take away reality, and all that is left is relativism, a belief that truth can be established simply by asserting it, that the self is all that exists – no, that *my*self is all that exists. The computing industry was built on the liberal belief in the individual as the only legitimate political entity, and virtual reality has, in some hands, been promoted as the ultimate embodiment of that principle. What better way of expressing your individualism than by creating your own, individual reality? Empowered by the personal computer, liberated by virtual reality, the individual becomes the God of his or her own universe. The sight of someone wearing a virtual reality headset is the ultimate image of solipsistic self-absorption, their movements and gestures meaningless to those left outside.

We have to look, then, at how virtual reality and artificial reality, the technology and culture, are changing public reality. Because the formidable might of commerce and in particular the computing industry have been deployed in defining reality, we need to look at what they mean to do with it.

Notes

1 Speech given at the anniversary of the *People's Paper* in Robert C. Tucker *The Marxist-Engels Reader* 2nd edition, London: Norton, 1978, p. 578.

2   Nina Eurman, 'Fake fat', *Health*, October 1987, p. 8.

3   Martin Bailey, 'Calling all fly fishermen', *Observer*, 3 November 1991, p. 3.

4   'Calorie-free carcinogen?, *New Scientist*, 3 December 1987, p. 25.

5   Andrea Gabor, 'Have your cake and eat it, too', *US News & World Report*, 23 May 1988, 104 (20) p. 58.

6   Jürgen Habermas, *Legitimation Crisis*, trans. Thomas McCarthy, London: Heinemann, 1976.

7   David Harvey, *The Condition of Postmodernity*, Oxford: Basil Blackwell, 1989, p. 262.

8   Umberto Eco, *Travels in Hyperreality*, London: Picador, 1987, p. 126.

9   Meckler, 'Virtual reality 91', Conference Forum, London, 5 June 1991.

10 Myron Krueger, *Artificial Reality*, Reading, Mass,: Addison-Wesley, 1983, p. xi.

11 This precise definition does not appear in William of Occam's surviving work, but it is the one most philosophers use. See, for example, Anthony Flew, *An Introduction to Western Philosophy: ideas and Argument from Plato to Sartre*, London: Thames and Hudson, 1978.

12 Roland Barthes, *Mythologies*, London: Paladin, 1973, p. 11.

13 Ibid., p. 130.

1
EUPHORIA

'I think this is one of the most important meetings ever held by human beings.' These words were spoken by Dr Timothy Leary, the Harvard academic and patron of the 1960s drug culture, who invited a generation to 'tune in, turn on, drop out', and may well have attended enough meetings held by non-human beings to be in a position to make the comparison. Nevertheless, the meeting he was referring to hardly seemed to justify such an extravagant billing. He may have invoked the ghosts of Plato, Max Planck, Marshall McLuhan and the Grateful Dead, declared the dawn of a new era, thanked and loved all the people who had brought it about, but it was still just an annual technical computer conference held in Dallas in August, 1990.

John Perry Barlow, a songwriter who once worked with the Grateful Dead, said in a talk given at the same conference that Leary was the opposite of the coalmine canary; he was one that 'starts jumping up and down talking about how great the air is in here'. In an equally memorable phrase, Barlow also described bullshit as 'the grease for the skids upon which we ride into the future'. Bullshit there was in abundance at that epochal meeting, and Leary the canary jumped around and twittered in excitement at the smell of it. So did everybody else. For this was a meeting of the computer revolutionaries and virtual realists, the creators of more than a mere technology or even a new stage of industrial development. These were the people who had, in Leary's words, initiated a new stage of human evolution.

The conference was SIGGRAPH, the annual computer graphics meeting run by the Special Interest Group, Graphics of the American Association of Computing Machinery (which is an association of computing companies, not, as its name suggests, of machines). Despite its pedestrian name and subject, SIGGRAPH is perhaps one of the most impressive annual industry events to be held in America. Its venues are the great conference halls that have become the landmark buildings of American cities, attracting tens of thousands of delegates and a celestial (by computer industry standards) cast of speakers.

I first attended SIGGRAPH in 1989, when it was held at the Hynes Convention Center in Boston, Massachusetts. Hynes is huge, a marble mausoleum, perhaps only too symbolic of the ambitions of its patron, the state's governor, Michael Dukakis, who the previous year had been defeated by George Bush in the presidential election. How could a computing industry conference, I wondered when I first walked into the center's majestic lobby, justify hiring a building with the proportions of a cathedral?

If anything, it turned out to be too small. The place thronged with busy delegates – an estimated 30,000 – all of them with pennants of identification stuck to their lapel badges: 'exhibitor', 'speaker', 'press', 'official', even 'pioneer'. Around 225 stalls – some the size of a suburban semi, and probably as expensive to build – filled cavernous exhibition halls measureless to man.

The awesome scale was partly explained by one of the more striking images to emerge from the event, one generated not by a computer but by Senator Al Gore, a member of the US Congress's science and technology committee. 'Information', he told a packed inaugural meeting by way of a pre-recorded video, 'is exploding by leaps and bounds.' And SIGGRAPH was evidence of this. The group's membership had exploded by leaps and bounds since its foundation in 1967, when it comprised a few programmers regarded by the rest of the industry as having a petulant preference for pretty pictures over the indecipherable code that made up serious software.

By the beginning of the 1990s, computer graphics were established not simply as an important part of the computing industry, worth an estimated $11 billion a year in the US alone and growing at an annual rate of 15 to 20 per cent. They had become a new medium, a potential successor, even, to television and print. This was the preoccupation of many of the panel discussions at the 1989 conference, as was the issue of whether or not computers could yet be used to produce 'art'. All in all, there was no mistaking a sense of destiny, that SIGGRAPH was becoming a meeting place for the future, where the key ideas of tomorrow would first surface.

And so it would seem to be, for it was at SIGGRAPH 89 that virtual reality made what might be regarded as its grand entrance. Though early forms of the idea and the technology had been knocking around the conference circuit for a few years prior to that (I first encountered it the previous year at a film festival in the Hague), this was the moment when it was ordained by the SIGGRAPH organizers as an idea worthy of serious discussion, which it duly received in a panel session entitled 'Virtual environments and interactivity: windows to the future'.

The star of the discussion was Jaron Lanier, the founder of a company called VPL Research. Lanier has come to symbolize the virtual reality industry. He represents its West Coast origins and values. He wears dreadlocks in his auburn hair; he is chubby; he talks fast and apparently in parallel, several lines of thought being expressed at once; he eschews suits and business formalities; he does not hold meetings, still less keep appointments, preferring spontaneous encounters.

At that first SIGGRAPH panel, he eloquently described the experience of virtual reality thus: 'It's very hard to describe if you haven't experienced it. But there is an experience when you are dreaming of all possibilities being there, that anything can happen, and it is just an open world where your mind is the only limitation. But the problem is that it is just you, you are all alone. And then when you wake up, you give up all that freedom. All of us suffered a terrible trauma as children that we've forgotten, where we had to accept the fact that we are physical beings and yet in the physical world where we have to do things, we are very limited. The thing that I think is so exciting about virtual reality is that it gives us this freedom again. It gives us this sense of being able to be who we are without limitation, for our imagination to become objective and shared with other people.'

On the exhibition floor, many people had the first opportunity to try this experience out for themselves. Two companies, a software house called Autodesk and VPL itself, were demonstrating virtual reality systems. For VPL a man sat on a swivel stool, wearing a pair of awkward goggles called 'Eyephones' which had tiny colour TV screens for eyepieces and a position sensor on the headband to detect head movements. He also wore a 'Dataglove', a single black silk glove with strands of optical fibre stitched into the fingers to detect finger movements, and another position sensor attached to the wrist to detect the overall position of the hand. Both goggles and glove were connected to a computer via long, heavy umbilical wires.

VPL's demonstration, entitled RB2 (short for 'reality built for two'), allowed Eyephone wearers to look into a crude, cartoon-like 'world' which they could move around by gesturing with their Datagloved hand (which itself would appear as a cartoon hand whenever they held it up in their field of view). As they turned their head or moved through the world, the scenery would change accordingly. Also, if someone else was wearing another Eyephone and Dataglove connected to the same system, the two wearers could meet them in this world, though their bodies would take the form of cartoon characters.

The Autodesk demonstration was called 'Cyberspace' – a word taken from William Gibson's book *Neuromancer* – and used standard personal computer equipment to produce the graphics, so was not quite as sophisticated as VPL's. However, it did have a neat demonstration of low-gravity squash. This entailed replacing the Dataglove with a standard squash racket with a tracking sensor fitted to the handle. The Eyephones revealed a primary coloured squash court in which floated a ball and a racket. As you moved the racket in your hand, so the virtual racket in the virtual court moved accordingly. You played simply by trying to hit the virtual ball with the virtual racket. Because the gravity was turned down so low (it was adjustable), the lightest tap would cause the ball to bounce wildly around the room. As a game, it had little to recommend it, as a demonstration, it was quite convincing, even though the movement of both the scenery and the racket lagged behind your body movements, rendering a very poor illusion of 'reality'.

But the freedom of dreams? A world without limitations? A place in which, as Lanier also claimed, people played tag by hiding inside each others' heads, in which they had even started flirting with each other? It was not the world I had experienced on the exhibition floor. I had experienced a crudely rendered, primary coloured series of badly coordinated images. I got none of the promised sensations of liberation or even disorientation, just frustration at the unresponsiveness of the equipment.

Of course, I was using one of the earliest systems developed. The first television and film systems were similarly crude. If anything, the technology's lack of refinement merely served to show that we were witnessing the creation of something new. However, Lanier's rhetoric was not about the future, it was about the present. This technological liberation from reality was, he claimed, already underway. 'However real the physical world is . . . the virtual world is exactly as real, and achieves the same status,' he said, 'but at the same time it also has this infinity of possibility.'

Lanier was by no means alone in making such claims for virtual reality. At that first SIGGRAPH meeting, not a single question was raised about the validity of his claims. It was accepted that available technology had shown how simulation could catch up with reality. Others, notably Scott Fisher of NASA, gave far more down-to-earth accounts of their research in the field. But the overall mood, partly set by the meeting's moderator, Coco Conn, who insisted that the proceedings be unstructured, was not one that matched the mood of the other panel discussions (which had titles such as 'Effective software systems for scientific data visualization'). If anything, it better reflected the language of the exhibition itself, where companies that year were promoting themselves with slogans such as 'the fastest route from imagination to reality'.

It was in this context that, the following year, SIGGRAPH decided to hold a two-part panel entitled 'Hip, hype, hope: the three faces of virtual worlds'. By this time, virtual reality had been identified as the 'Big Idea' of the 1990s and become a source of intense media interest and industry debate. Reflecting this, Bob Jacobson, the meeting's moderator, selected a multi-faceted panel. Besides Timothy Leary, the shaman, and John Perry Barlow, the minstrel, there was Esther Dyson, the editor of an industry newsletter and a voice of caution, William Bricken of the University of Washington's Human Interface Technology Laboratory and Warren Robinett of the University of North Carolina, who were the hard-headed technologists, and, inevitably, Jaron Lanier. Nolan Bushnell was also advertised to appear, but failed to turn up.

The selection of Bushnell was, despite his absence, one that to a large extent reflected virtual reality's emerging 'Big Idea' identity.

While working as an engineer for the Californian recording equipment and tape manufacturer Ampex, Bushnell set about designing coin-operated computer games. His first effort flopped, but his second, a computer version of ping-pong he called Pong, was a huge success. He left Ampex and set up a new company to sell the game, which he named after the Japanese expression used in the game Go which means 'I am going to attack you', Atari.[1] Atari was to become the incubator of virtual reality. It set up a research laboratory at its Sunnyvale headquarters which was to employ Jaron Lanier, William Bricken, Warren Robinett, Scott Fisher (the NASA researcher not on the hip, hype, hope panel, but in the audience and hailed by Leary as a 'beloved digital engineer') and Brenda Laurel (another of Leary's beloved digital engineers).

There must have been something in the water at Sunnyvale, because those who worked there in the early 1980s went on to shape the agenda of personal computing for the following decade. Some, notably Alan Kay, promoted the idea of the computer being some sort of mental amplification tool at companies like Xerox and Apple, and others went on to discover virtual reality. What encouraged them was not just their enthusiasm for developing the computer game as model human/computer interaction; it was the company's role as the creator of a new sort of research institution. Buoyed by the international impact of the products that had made them famous, Silicon Valley companies like Atari and Apple had acquired a taste for technological revolution. It was the job of their flush research departments to keep the revolution going. Rather than simply produce new variations on the same Pong game, Atari (which had been bought by Warner Communications in 1976) aspired to repeat Bushnell's original insight. It wanted to turn computer games into a whole new medium, a competitor to television and film. People like Fisher were appointed specifically because they had worked on academic research teams that had already addressed this issue, specifically in the Architectural Machine Group (known as Arch Mac) at MIT, set up by Nicholas Negroponte, who later founded the MIT Media Lab. Arch Mac was responsible for developing what was to be regarded as a ground-breaking experiment, the Aspen Movie Map. The concept was simple, to have a movie that shows a drive through the town of Aspen, Colorado, with the difference that the viewer, by pointing at the screen, could alter the course of the drive. This was done by filming a number of alternative drives through Aspen, and the using a computer to control the order they are shown, the order being determined by the viewer's chosen points of interest. The Movie Map demonstrated the feasibility of creating a cross between passive media like TV and active media like games.

Virtual reality, therefore, came from a research environment which had already set itself the task of challenging television – which, as it turned out, computer games were to do, though it was Nintendo, not Atari, that was to do it. What made this challenge unique, and what gave it, perhaps, a uniquely adventurous and aggressive thrust, was the fact that it came from a commercially funded research environment, one that gave its researchers quasi-academic freedoms to indulge in 'blue sky' work, rather than product-orientated development. As a result, it has been subject to none of the usual academic controls: there has been no coordinated experimental work to test its hypotheses confirmed by other laboratories, and the notable papers have tended to appear in magazines like *Scientific American* rather than in journals like *Science* or *Nature*.

The significance of virtual reality, then, is to be judged more according to news than formal scientific values. It is a set of ideas that is liberated from the control of any particular institution and which therefore can be expressed using none of the conservative, inhibited language associated with academia. The virtual reality community formed a loose-knit but like-minded group of entrepreneur technologists – the key members of which have known each other for much of their careers. They were more like an artistic movement, complete with, as we shall see, its own rhetoric and political agenda. It was this community that gave virtual reality its spirit and identity, and which you really had to belong to in order to engage in VR research. Much work that seemed to be relevant to the development of VR, particularly in the area of simulation, was simply not counted as being part of the canon of research because the people doing the work were not part of the movement. It was not that the movement set out to exclude anyone – most of its members were only too pleased to welcome newcomers – it was that it had defined the term and so had by default become the custodian of the sacred scrolls. And it was the content of those scrolls that a theatre-full of delegates and press had come to learn at the hip, hype, hope panel of SIGGRAPH.

The meeting began with an air of unprecedented expectation: some magic was about to be wrought by the computer wizards, and this was to be the moment of epiphany. Bob Jacobson introduced the speakers with a speech that placed them firmly in the context of paradigm shifts and pioneers. He admitted that 'the technology is somewhat primitive' but promised that 'we are bringing our best resources to bear, to make it happen more quickly, to serve you and to serve our ultimate beneficiaries, the people who see the things that we produce, who experience and learn by them'. This cosy little constituency of 'you' and the 'people', all of whom were in need of whatever 'we' were so inevitably going to deliver, seemed somewhat to undermine the spirit of the panel's sceptical title.

Nevertheless, that was the agenda, and Warren Robinett set about examining just exactly what 'we' were up to. He did voice some scepticism about the term 'virtual reality', which he described as a 'cute little oxymoron', preferring to use 'synthetic experience' instead. But he accepted that it was unlikely to go away, and proceeded to outline a set of adventurous but practical experiments that conformed to the VR orthodoxy. One was for 'X-ray vision'. He imagined a pair of half-mirrored virtual reality goggles where the virtual image is superimposed on what the wearer of the goggles sees through the lenses. He illustrated the idea with a picture of a doctor examining a scan of a pregnant woman's womb, with the image of the foetus, built up using information from a coventional ultrasound scan, superimposed in the appropriate position on her abdomen. Using this system, the doctor could look around the abdomen, with the computer changing the view of the foetus projected onto the goggles' lenses accordingly.

Having considered some of the applications, Robinett contemplated some of the imponderables. What, he wondered, do 'imperceptible things' look like? If, for example, you have a three-dimensional model of a molecule, and want to inspect it using a pair of goggles, how do you visualize it? A molecule is invisibly small, so what does it look like? Those with any sort of image of a molecule tend to think of it as a series of coloured ping-pong balls connected together by cocktail sticks. But real molecules do not, of course, look anything like that. The ping-pong ball model is a very simplified illustration of the structure of molecules: atoms are no more like ping- pong balls than they are like footballs, or rain-clouds, or solar systems.

After Robinett's sober, sensible assessment came Jaron Lanier's. He introduced his talk with an exclamatory 'yowzza' and some reflections on 'h' words other than hip, hype and hope that characterized the 'three faces of virtual worlds' (there were hoops, 'which we have to go through', and happenings). This was a strategy that seemed to entail leaving the first three faces conveniently uninspected. There then followed one of Lanier's famously unstructured peregrinations through a loose alliance of ideas roughly linked to computer simulation and human perception. It was a bravura performance, the sort that made him the centre of so much press attention – indeed, VR has been described as a form of total immersion in media, and, in Lanier's case, this was proving to be almost literally true. While restating his belief that virtual reality revives our childhood powers of fantasy, he warned that 'this stuff is getting amplified by the media'. A little later, he counselled that 'there is a really serious danger of expectations being raised too high', and that 'the level of interest is just a little premature'.

It was easy to sympathize with the man's diminishing desire for media attention. He had received it in buckets the moment he announced his RB2 system. The *New York Times* had featured him on its front page as far back as April 1989, describing Lanier as a 'guru'.[2] Any desire he may have had for publicity seemed genuinely to have been satisfied by the acres of coverage that were to follow. Nevertheless, his faith in the technology was undiminished. Though the systems demonstrated at the 1990 SIGGRAPH exhibition seemed little better than those of the previous year, he suggested that it would take only a few more years for their limitations to be overcome.

Lanier was followed by William Bricken, who provided what must be regarded as a key statement of the field's academic ambitions, as well as some spectacular jargon. He identified what he called a 'research suite' of technologies that needed further study: 'behaviour transducers', 'inclusive computation' and 'physiological psychology'.

What was, perhaps, the most significant aspect of Bricken's 'suite' was its emphasis on psychology. This properly reflected one of the main preoccupations of virtual reality, at least as defined by those sitting on the SIGGRAPH panel. Bricken coined a phrase that has since become one of the slogans of VR: 'psychology is the physics of virtual reality'. It is hard to know what is meant by this. Perhaps it means that, just as physics is about revealing the laws that govern the physical realm, so psychology is about revealing the laws that govern the mental or imaginary realm. Virtual reality, then, is about discovering this world as it is determined by these laws – and is based on the assumption that such laws are in some sense mathematical or, more precisely, 'computable' a term we shall return to later.

Bricken, however, spoke for the entire assembly of virtual reality enthusiasts gathered at that meeting when he expressed this 'physics' in terms that go way beyond the conventional language of science. 'We are talking about much more than reality,' he said. 'VR is not a physical simulation. In fact, when you are buildings systems in order to get a physical simulation, what you do is take your VR and add contraints to it. Throw away the constraints and you're in something that's a bigger space than physical. We have new freedoms, we have new things to learn.' Echoing Marshall McLuhan, he argued that each new technology is initially understood in terms of its predecessors: the car was the horseless carriage, television was radio with pictures, and so virtual reality is currently seen as physical reality with a few extra features, 'and reality', Bricken triumphantly concluded, 'is in the eye of the beholder'.

If ever there was a sentence that characterized the gospel of the most eagerly-awaited panellist, Bricken's conclusion was it. In 1970, a US federal judge described Timothy Leary as 'the most dangerous man in the world'. Dangerous or not, he is one of the most beguiling men in the world, a notorious, fabulous shaman, showman, huckster, hawker, thinker, doer. His SIGGRAPH talk honoured his oratorical reputation, making the audience feel that virtual reality truly was the most important idea of the late twentieth century, and that, if not the most important ever held by human beings, this was certainly a meeting worth its entrance fee.

Leary's early life is the story more of significant departures than meetings. He was born in 1920 to a fun-loving army dentist (an unusual combination) who was expecting to inherit a fortune from his parents, reputed to be the richest Irish Catholic family in west Massachusetts,[3] and a god-fearing mother of far less glamorous agricultural stock. When Leary's father eventually inherited, all he got was a few thousand dollars. In 1934, he gave some money to his wife, a little to his son, and disappeared for ever.

Leary described his father as a 'model of the loner, a disdainer of the conventional way' who 'dropped out, followed the ancient Hibernian practice of getting in the wind, escaping the priest-run village, heading for the far-off land, like one of the wild geese of Irish legend.'[4] Mother Goose was very different, a model of respectability and piousness, and Timothy, inevitably, found himself identifying with the absent and wild, rather than the present and domestic, parent. He wanted to fly with the geese, not remain stuck in the mud of his mother's family's farm.

And he flew, though through skies that were to prove sometimes violently turbulent. He briefly joined the army, only to be drummed out for bad behaviour. He went to the University of Alabama, where he discovered psychology, James Joyce's *Ulysses* and was himself discovered one night in the women's dormitory, for which he was expelled. During the war he rejoined the army as a psychologist, and after he went to Berkeley where he was awarded a doctorate in clinical psychology. After that, he joined the Oakland's Kaiser Hospital as director of research, married, had two kids, and settled down to what could have become a life of domestic, bourgeois respectability.

Then, when he was 35, his wife committed suicide. The tragedy prompted Leary to give up his job and take off to Europe with his children. There he wandered aimlessly between lecturing jobs, fetching up in Florence where an old Berkeley friend, Frank Barron, came to see him with a bottle of whisky and news of some mushrooms.

Maybe this meeting was not the most important ever held by human beings either, but it was one that set in motion a train of events that were to have a profound effect not just over the subsequent 1960s 'counter cultural' revolution, but over the subject of the meeting Leary was to attend 32 years later. Before flying out to Europe, Barron had tried some mushrooms he had read about in *Life* magazine, teonanacatl (*psilocybe mexicana*), which had a narcotic effect. Intrigued by their possible application to psychology, Barron had tracked down two scientists at the National University in Mexico City who had plentiful supply of the mushrooms and were willing to share them with fellow researchers.

Barron also happened to mention to Leary the imminent visit to Florence of David McClelland, the director of Harvard's Personality Clinic. It was this piece of news, rather than Barron's discovery of magic mushrooms, that he was to act upon. On the strength of a subsequent meeting he had with McClelland, he was offered a lectureship at America's most prestigious university.

It was at Harvard, more precisely the Center for Personality Research, that Leary was to embark on the experiments that were to turn him into one of the key figures of the 1960s. Using psilocybin, the narcotic ingredient of the Mexican mushrooms, Leary, together with Barron, embarked on a series of experiments designed to discover the psychoactive effects of such drugs, and the otherwise hidden regions of the mind they seemed to reveal.

At the time, such experiments were regarded as a perfectly respectable academic pursuit, part of Leary's investigation into what was somewhat extravagantly called 'existential transactional psychology', which treated behaviour as a game, with different roles being the product of different sets of rules. However, as the experiments continued, and as Leary himself became a more and more active participant in them, what had begun as a putative programme of serious scientific research began to look more and more like very unscientific fun. It also showed alarming signs of developing into a full-blown social and cultural revolt.

To an existential transactional psychologist such as Leary, it was all a matter of game-playing anyway; he was simply choosing to play according to a different set of rules. The academic researcher game was being substituted by the pioneer-cum-guru game, and it was obvious which he preferred playing.

But it was a political game, too. With LSD taking over from psilocybin as the focus of interest, the stakes were becoming too high for the academic authorities, and the attempts by the gentlemen to spoil the fun for the players was seen by Leary and his like as breaking the rules of a much bigger game, the game of civil rights. Leary even saw his experiments as the expression of a new civil right. In a speech given to the Congress in 1941, the great liberal president Franklin D. Roosevelt outlined four freedoms that he saw as fundamental to all human affairs: freedom of speech and expression, freedom of worship, freedom from want, freedom from fear. Leary proposed a 'fifth freedom': the freedom to do whatever one wished with one's consciousness. He even proposed an amendment to the constitution preventing Congress from passing any law that would limit the individual's right to seek an expanded consciousness.[5]

'There is this 30-foot tall palm tree with diamond coconuts. The boy is a 12-year-old Polynesian with a blue bikini. He's doing a double back-flip into a swimming pool filled with marshmallow fluff.' This could be Leary describing his Harvard experiments, but it was his description of a virtual reality presented to the SIGGRAPH panel. The similarity of the language, however, is no coincidence. He clearly saw virtual reality as the continuation of his exploration of consciousness conducted by other means.

The link with LSD was established early on in the history of the virtual reality. It was, perhaps, inevitable, given the interest of Leary. It certainly gave the idea news value for journalists, who at every opportunity tried to lure a very wary Leary and Lanier into describing VR as a new form of hallucinogen. But the connection with drugs was a symptom more than a cause of the relationship between Leary and VR. The two were inextricably entwined by their histories.

One of the earliest published references to virtual reality came in a book written by Stewart Brand entitled *The Media Lab: Inventing the Future at MIT*. The MIT of the title was, of course, the Massachusetts Institute of Technology; the Media Lab (motto: 'Our charter is to invent and creatively exploit new media for human well-being and individual satisfaction'[6] – note the *Star Trek*-like half-split infinitive) was a laboratory set up as the world's first research centre devoted to the technologies of the media age. 'The term "virtual reality" tickled a different giggle in me every time I heard it', wrote Brand. 'I began to realize why artificial intelligence and robotics and [computer-generated] animation have become such fundamental science. The researchers are reinventing the world from scratch. . . . We are banging our heads in the playpens of another level of understanding.'[7]

Stewart Brand established himself in the 1970s as one of the official chroniclers of the computer revolution. He was also a key participant in Leary's 1960s psychedelic revolution. Having studied biology at Stanford University in California, he became one of the 'Merry Pranksters', the group of acid voyagers that were at the vanguard of Leary's psychedelic revolution and who drove around in Ken Kesey's Dayglo bus. Brand organized one of the key events of the LSD era, the 'Trips Festival' held in San Francisco in January, 1966. It was to be the last of a series of events dreamed up by Kesey called 'acid tests', group acid-dropping sessions aimed at exploring the idea of a 'group mind', a state of collective psychic intimacy that caused individual minds to melt into one single, seamless consciousness. It was at the second acid test, held after a Rolling Stones concert in San Jose, that this aim came closest to fulfilment. Attracted by handbills distributed by Pranksters at the end of the concert, a group of Stones fans convened at the rambling house of a local eccentric and began taking LSD to the accompaniment of a Palo Alto group called the Warlocks – who were later to change their name to the Grateful Dead, the band for which another SIGGRAPH delegate, John Perry Barlow, wrote lyrics.

Brand clearly saw the acid test as a new type of public event, one that could, given an appropriate dose of organizational and entrepreneurial flair, rival a rock concert. Brand was not interested in Roneoed handbills; he hired an advertising executive as publicist and – hard though it may now be to believe – set about attracting business sponsors. Brand's commercial pragmatism and boy scout enthusiasm resulted in a sort of huge village fete, one that attracted an estimated 10,000 people and perhaps, though this goes unrecorded, a profit. It was so successful that a New York promoter reportedly wanted to book the acid test for Madison Square Garden (which the Pranksters would only agree to, according to the Trips Festival publicist, if it was renamed the Madison Hip Garden).[8]

In the event, Kesey was arrested for possession of marijuana soon after (while he was on the roof of Brand's apartment, as it happens) and LSD was outlawed, turning the drug movement into one with a less relaxed agenda. In 1968, Brand went on the edit the *Whole Earth Catalog*, a definitive resource book for the ecologically minded, providing information on crafts and kind living for the counter culturalists surfacing from underground

movement. The *Whole Earth Catalog* was radical in its design as
well as its content, loose in its layout, displaying no particular
order or structure. This was, of course, intended to reflect its
publishers' free-thinking philosophy, a reluctance to give one
piece of information primacy over another on the grounds that it
was more democratic to let the readers decide for themselves
what was important.

As the 1960s drew to a close, California was already a centre of
the sort of high-technology companies and prestigious academic
research centres that were to make up Silicon Valley, and a
number of those who had been lured there by the prospect of
revolution began to see the computer as the means of achieving it.
And it was, ironically, the philosophy expressed by the low-tech
Catalog that helped foster this belief.

The connection of high and low tech can, in part, be traced to
the influence of Buckminster Fuller, an architect-cum-inventor-
cum-engineer who preached a gospel of technological humanism
that resonated with a generation more familiar with the technological oppression of war. Fuller claimed that there were quite
sufficient resources to serve all humanity, the only problem lay in
their deployment. As a response to this he developed the concept
of Dymaxion, a contraction of the words 'dynamic', 'maximum'
and 'ion' that to him summarized the need to develop resource-
efficient, self-sustaining technologies. He developed the Dymaxion
ideal by inventing a host of devices, from lightweight homes
through streamlined cars to the geodesic dome that, thanks to
Disney's EPCOT theme park, has since become a symbol of
futurism.

He also produced what he called the World Game. Based on his
Dymaxion Air-Ocean World Map, which was the size of a
basketball court, it entailed trying to distribute world resources in
a way that ensured that everyone would 'win'. This very whole
earth entertainment drew the attention of a number Californian
programmers and philosophers, who managed to implement a
version of it using a computer, perhaps in the process creating
the world's first world simulator and setting a precedent that was
to make the idea of virtual worlds more accessible – at least, to
California's free-thinking programmers.

By the early 1970s, California, and particularly the San Francisco
Bay Area, was filling up with research programmers, electronics

engineers and broad-minded liberal humanist hangers-on with a desire to create a whole earth, free-thinking environment that captured Brand's brand of entrepreneurial counter culture. Bob Albrecht, a former computer company employee, and a teacher, LeRoy Finkel, were typical of the times. They founded a publishing company called Dymax, in honour of Fuller, with a mission of computer enlightenment. In October, 1972, they launched a tabloid newssheet entitled *People's Computer Company* which, in its launch issue published in October, 1972, carried the following message:

'Computers are mostly
Used against people instead of for people
Used to control people instead of to FREE them
Time to change all that –
We need a . . .
People's Computer Company'[9]

Two years later the message found equally forceful expression in a book with aspirations to become the *Communist Manifesto* of computer culture, a book whose author was later hailed by Leary with characteristic hyperbole: 'Well, up on some magnificent Himalayan mountaintop, next to the cerebral left or right lobe, where a few men and women get to look down over human affairs and can predict what's going to happen – we put Galileo, Tom Jefferson, Tom Paine, and Buckminster Fuller. There for 13 years, and maybe a thousand, has been Ted Nelson.'[10]

Modelled on the *Whole Earth Catalog*, Ted Nelson's *Computer Lib* was a resounding, finger-twirling creed of technological liberation. In a second edition published in 1987, Stewart Brand described its unashamedly political mission: '"Revolution" is an overvalued word these days, especially around new technology, most especially around computer technology. For the coming of personal computers, however, it's the right word, because the people who made the event happen saw themselves as revolution-aries and acted accordingly. In motivation, style, and global effect, they fought and won a Second American Revolution. Oh, was it American.'

The son of a Hollywood star and a movie director, the product of expensive schools and art colleges, the writer in 1957 of a rock musical, Nelson was a perfect candidate for the role of revolutionary.

He was charismatic, confident and clever. He brought a spirit of adventure to the notoriously beige, conservative world of corporate computing.

As a political and moral tract, *Computer Lib* is as crude as its title would suggest. Nelson used placard language, right down to the use of capitals for emphasis: 'THIS BOOK IS FOR PERSONAL FREEDOM AND AGAINST RESTRICTION AND COERCION' (and against baby battering too, presumably), 'COMPUTER POWER TO THE PEOPLE! DOWN WITH CYBERCRUD!' Its message was simply that everyone should learn how to use computers because computers are much more interesting and important than everyone assumes them to be. In the service of this cause, copious advice on how to back up data and acquire basic programming skills was interspersed with calls for computers to be made available to everybody who wants them – even suggesting that families club together to buy minicomputers.

The book had its baddies, too, the worst of which was IBM. In the early 1970s, computing was monopolized by IBM. It represented the antithesis of technological liberalism: it was notoriously possessive of both its technology and market; its customers were from the military-industrial complex that every free-thinking Californian radical knew to have fuelled the Vietnam War; it promoted ignorance by making its computers complicated and by intentionally failing to document adequately how they worked; its technology was based on the principle of centralization, comprising enormous mainframes offering computing power only to those who management chose to give the privilege of access; its technology was obsolete, designed to hold back the fulfilment of the Nelsonian dream of universal computing.

The white stetson, meanwhile, was worn by Digitial Equipment Corporation, DEC. Described by Nelson as 'The computer fan's computer company', the highest possible accolade, DEC was praised for building small computers and supplying them with full documentation. The company was also praised for building computers that were 'designed *by* programmers, *for* programmers'. This may sound like an award for exclusivity rather than universality. But programmers were the solder-gun toting cowboys of the technological frontier. They were the pioneers, and Nelson's dream was that the wilderness they had tamed would one day accommodate us all.

Two years after the publication of *Computer Lib*, Nelson was photographed posing on a rock dressed as an ape holding an Altair 8800 computer triumphantly aloft. He showed the resulting slide to a group of local government officials in Atlanta to the accompaniament of the music from *2001: A Space Odyssey* and while proclaiming that 'COMPUTERS BELONG TO ALL MAN-KIND'. The computer revolution was underway, and as promoted by Nelson, it was a moment that even a bunch of beguiled civic dignitaries had to applaud.[11]

* * *

The Altair 8800 was an electronics kit that first went on sale in December 1974, just months after Nelson's book was published. It is claimed to be the world's first microcomputer offered for sale. It provided, however, meagre computing resources, cost several thousand dollars in any operable configuration, and served no useful purpose. But it was a start. It also inspired a couple of young enthusiasts who hung out at the People's Computer Company to have a go at designing a computer of their own. They were Steven Jobs and Steve Wozniak, and the computer they built was the Apple, in name as well as design, the prototype of the new era of personal computing and benign whole earth business values. Apple, which has credited itself with the invention of personal computing, was the company that promised to turn the Nelsonian dream of computing power to the people into a practical possibility. In 1984 it launched the machine that consolidated its international success, the Macintosh. It hired Ridley Scott, the director of *Blade Runner*, to direct an advertisement that did not even bother to show the product, but rather depicted the announcement of the new machine as a moment of liberation when the totalitarianism of the mainframe era (in other words, the IBM era) would end and the individualism of the personal computer era, represented by a white-teeshirted woman athlete pursued by secret police, would begin.

However, as Apple's power has grown, so its counter-cultural credibility has diminished. By the late-1980s, it was being run by John Sculley a former Pepsi executive who, despite an enthusiasm for informality at Apple conventions and a sometimes visionary oratorical style, was suspected of wearing a corporate suit beneath his teeshirt. Like the industry that it had done so much to influence, Apple had helped turn revolution from a political movement into an advertising slogan. Even more sinister from the point of view of the computer liberationists, the term 'personal computer' had actually been adopted by IBM as the name for its microcomputer products. By the end of the 1980s, describing a product as 'Personal Computer compatible' actually meant that it was functionally identical to the IBM PC. The forces of tyranny had, in short, succeeded not so much in quashing the revolutionary cause as taking over the running of it.

It was, signficantly, during this era of corporate takeover that the term 'hacker' began to pass into everyday language. Its meaning was ambiguous. It was used approvingly by those who were identified as hackers, and pejoratively by almost everyone else. To both, hackers were obsessives. But to the hackers themselves, they were obsessive about exploring the ways computers work. They were explorers of the information frontier, followers of what has become known as the 'hacker ethic', a creed defined by Steven Levy, in his book on hacker culture, as a demand that 'access to computers – and anything which might teach you something about the way the world works – should be unlimited and total'.[12] They acknowledged no restrictions to these rights of access, and were only too happy to breach any barrier put in their way. In contrast, non-hackers saw hackers as saboteurs, unscrupulous, reclusive, socially underdeveloped adolescents who casually broke into other people's computer systems in order to corrupt or steal the information they contain.

In the mid-1960s a student at the Massachusetts Insitute of Technology founded a contest to find the ugliest man on Campus.[13] It was not designed to be any sort of vendetta: people were actually expected to offer themselves for candidacy. In *The Second Self*, Sherry Turkle's book on the psychology of computers, this contest is interpreted as an expression of the engineer's sense of alienation, of being a casualty of the 'severed connection between science and sensuality'. This ironic self-degradation is a feature of hacker culture that denigrates the body but worships the machine. To a hacker, however, a machine is not what the non-hacker thinks it is. It is a machine in a computational, technical sense, an abstract system that runs according to the laws of mathematics. Hackers are, in other words, inhabitants of the abstract domain, and it is that domain to which they truly belong, not the ugly, mundane one of physical hardware.

No one better symbolizes the hacker's alien status than one of its most important and eloquent champions, Richard Stallman, who descibes himself as 'the last of the true hackers'.[14] Stallman was a researcher at MIT's Artificial Intelligence Laboratory, and a true militant of the cause. 'People will program for love', he once proclaimed. Not for a generous remuneration package, not for prestige, not even for an eternity of mind-mouldering servitude, but for love.

Stallman conducted his campaign for what he called 'programming freedom' from a small, dark cell in MIT's famous Tech Square tower, seen by many as the Eden of hacker culture. His room seemed to be his complete universe – indeed, little short of a big bang could have generated such a chaotic arrangement of clothing, bedding, books and papers. The only signs of order were a program listing displayed on a large VDU and a young, neat amanuensis who sat next to him, keying his bubbling stream of consciousness into the computer as he picked the dead skin from between his toes.

Stallman is a shock to behold in the age of aerobics. He looks like a pale-skinned, soft-bodied dweller of cavepools, ill adapted to the world's glare and noise. What, however, he may lack in muscle tone he makes up for in strength of spirit. In the face of the intimidating forces of commerce, he has fought resolutely to prevent what he sees as the enclosure of information. 'In the time that I've worked as a programmer', he has said, 'I've watched the field change from one of cooperating and sharing, where people could reuse previous work in any useful way to advance the state of the art, to one in which cooperation is largely forbidden by the owners of software.' This is, he thinks, part of a wider movement: 'There is a trend towards more control of what people can do with information.'[15]

Copyright, the main legal instrument for controlling information, was originally designed for authors to protect their creative effort. However, with the introduction of cheap copying technology – tape recorders, photocopiers, personal computers – it has increasingly been adopted by publishers to protect their profits. 'It is no longer possible to enforce any kind of intellectual property rights without a heavy hand', said Stallman. This was exemplified by the decision made by companies like Apple, Lotus and Ashton-Tate – companies that were themselves all participants in as well as beneficiaries of the personal computer revolution – to resort to law in an attempt to stop other companies using the 'look and feel' of their products. At stake was whether or not a program's 'user interface', the means by which the user manipulates it, can be copyrighted. If it can, argued Stallman, so can a typewriter's 'user interface', forcing each typewriter manufacturer to come out with a different arrangement of the keys, and so can a car's, forcing each car company to use different alternatives to the steering wheel. The fact that companies are even attempting to prosecute such claims shows how commercially important the control of information has become. Information owners have become dangerously possessive, according to Stallman: 'There's a tendency to destructive competition, where instead of trying to run faster yourself, you try to trip up everyone else.'

Stallman's reaction to this trend provides a case study in positive action for radical causes. To undermine increasing attempts by other departments within MIT to secure their computers with password systems, he rewrote the relevant parts of their administrative software so that as users typed in their password it would be displayed on the computer system's main VDU. He helped found the League of Programming Freedom and the Free Software Foundation to protest at the spreading commercial control of information and to promote the collaborative development of public domain software.

Such efforts, however, look increasingly futile in the face of the growing power of the personal computer industry. The slogan of freedom of information is becoming difficult to understand, let alone implement, in an industry that earns billions of dollars from it. Apple may have continued to pride itself on its free-wheeling, free-thinking corporate culture, but success may have gone to its head. Dismayed hackers began to accuse it of using the same technologically monopolistic practices that IBM had used – even that it was getting worse than IBM, which was at least allowing other companies to copy and develop some of the technology used in its Personal Computer (though, it should be added, that IBM did not allow this to happen because of its liberal principles; if it had not done so it stood a serious risk of prosecution under America's anti-monopoly laws). Under its Wall Street approved management, Apple was being accused of betrayal. It had used one Orwellian metaphor in the 1984 advertisement announcing the launch of the Macintosh. Another, that of Napoleon in *Animal Farm*, was, people like Stallman argued, becoming alarmingly relevant to its subsequent behaviour.

Had there, then, been a computer counter-revolution that had simply left a new set of plutocratic pigs in control? Certainly, notwithstanding its counter-cultural origins and the Californian location of its heartland, the computer industry emerged from the experience as politically conservative as any other, asserting a robust, red-blooded free-market philosophy that saw commercial competition as the only legitimate form of power struggle, and lawsuits as the only significant form of protest.

But what the virtual reality phenomenon has revealed is a somewhat complicated pattern of political responses. For people like Stallman, the hacker ethic represents a form of subversion that refuses to accept commercial control. He does not see the personal computer as a tool of freedom at all, since it has failed to change the commercial power structures that control information technology. For people like Brand, however, a very different reaction has emerged, one that sees the personal computer revolution as having wrought fundamental political change that simply needs further technological progress and some political encouragement to fulfill its promise. After being funded for two decades following the war by the American military, a 'funny thing happened' to computer science, wrote Brand in 1989:[16] 'Computing power dispersed. . . . Thanks to the deliberate grass-roots revolt of the creators of personal computers and the lavish cleverness of the makers of consumer electronics, the bit business began to be taken over by citizens and customers.' 'The personal computer revolution has vastly empowered the individual', announced *Reality Hacker*, the magazine that later mutated into *Mondo 2000* and was regarded by many in the virtual reality community as their philosophical journal.

But one has to ask: in what way? Which individuals are vastly empowered?

The example usually given of computerized people power is desktop publishing. Using a computer, the right software and a printer based on photocopying technology, together costing around half the price of a normal family car, it is possible to produce publications with professional-standard layouts and near-typesetting quality print quickly and relatively easily. This encouraged a welter of speculation – much of it fuelled by Apple, whose Macintosh first made its market impact as the main component of a desktop publishing system – about readers becoming publishers, of closing the loop between supply and demand, so that people no longer had to rely on the judgment of large, distant and disinterested corporations to supply their information. With the development of what became known as 'desktop video' alongside computer graphics and cheap video cameras, the same was anticipated for that most powerful and dominant of all media, television.

Brand's form of technological liberalism reached a peak in the mid-1980s, when the technology was beginning to deliver the goods. And it was at this moment that virtual reality was born. It represented the ultimate form of individual empowerment, the destiny of the revolution: it gave each individual command over his or her own universe. In the feudal era, lords could move villages and hills to landscape themselves the environment they wanted. In the personal computer era everyone will be able to do the same simply by donning their goggles and running a reality editor. Because the cost of the technology would inevitably fall, it would offer a truly democratic form of empowerment, available to anyone with the price of a hi-fi in their credit limit, including – Leary promised at the SIGGRAPH conference, the 14-year-old inner-city kid.

Could this ever be true, or is it another dump of John Perry Barlow's scented bullshit? Is it pure political rhetoric or a genuine technological possibility? It is hard to tell. The idea that desktop publishing liberated the press is highly suspect; there is little evidence of it actually happening in the fundamental sense of breaking down the traditional role of the publisher. It is also suspect for the simpler reason that samizdat and other 'unofficial', homebrew forms of publishing were achieved without computers – printing has, in fact, always been a cheap, accessible method of distributing information; that is what accounts for its enormous success. All that desktop publishing seems to have done is provide publishers with productivity gains.

The same argument can be applied to dekstop video, the concept that 'Prime time becomes my time', as Nicholas Negroponte, the director of MIT's Media Lab, put it. Certainly, cable TV as well as home video could, in combination with computers capable of manipulating video images, be used to create a sort of Me TV channel, one that reflects my individual interests. But the experience of home cine – surely itself a paradigm of liberating personal media – shows that it will take more than technology to challenge the position of media owners like Rupert Murdoch. Magazines for amateur film makers explicitly invited their readers to ape Hollywood, to create a sort of domestic studio system that would deliver them with films that reflected their lives. Hollywood survived. If anything, films have become less and less personal, more and more the product of an industrial system operating in an international market.

Television, it could be argued, is different. In America, cable TV has seriously threatened the power of the national networks. Perhaps this marks the beginning of a fundamental shift in the structure of the television market. But, even with high-technology cable systems offerings hundreds of channels and unlimited access, the most likely outcome is the development of TV more along the lines of newspaper and magazine publishing, where there is a mixed economy of national, local and special-interest media. Cable, like desktop publishing, has done little to change the underlying structure of media production.

To question the assumptions of the computer revolutionaries is not to deny the importance of the computer revolution itself. Its social and cultural impact has been immense. But the idea of it bringing about a sort of post-industrial perestroika does need to be examined carefully. To most outside the industry, the personal computer is a white-collar toy, an accessory for those already plugged into the power circuits. It is easy to forget that bullshit like 'the fastest route from imagination to reality' is not really for greasing the skids upon which we ride into the future, it is for greasing up to potential customers.

* * *

In October 1990, two months after SIGGRAPH, most of the hip, hype, hope panelists tripped off to San Francisco for the Cyberthon. It was to be the computer revolution equivalent of the Trips Festival held there 24 years earlier. It was even organized by

Stewart Brand, who once again invited the participation of local businesses which once again offended the radical sensibilities of the counter-cultural diehards – including Lanier and Barlow, who at one point threatened to boycott the event.

The link with the old acid test days was quite explicit. One delegate was even reported to have exhorted the assembly to recapture the 'conspiratorial ambiance of the dope-dealing past'.[17] However, it was an event that seemed to mark not the recovery of the spirit of the 1960s so much as the end of it. The first sign that virtual reality was passing a point of no return occurred at SIGGRAPH in the closing stages of hip, hype, hope. In a dramatic podium putsch, the computer artist Myron Krueger shoved a protesting Bob Jacobson to one side to announce his commitment to the idea, having coined the term artificial reality in the 1970s. 'I think Jaron [Lanier] was eight years old when I started working on this stuff', he said. Lanier wearily replied in terms that seemed calculated to damn Krueger with extravagant praise: 'This is Myron Krueger, who is a wonderful pioneer. I've been influenced by him, I respect him enormously, and I appreciate him incredibly.'

By the time Lanier was complaining about the suits at the Cyberthon, his company, VPL, had become firmly established as the leading supplier of virtual reality 'tools', not simply headsets and gloves, but suites of software that would enable users to build 'worlds' for themselves.

Meanwhile, Autodesk, another of the pioneering companies, was trying to take possession of the term Cyberspace, only being dissuaded from doing so when William Gibson, who coined it, threatened to sue. (Ted Nelson, now working for Autodesk, was more successful; 'Xanadu', which he had originally adopted in honour of Coleridge's poem, 'Kubla Khan', was established as a trademark for his 'hypertext' system). Such a rush to lay claim to intellectual property rights was a clear sign that the forces of commerce were firmly in charge, and no amount of clamour about returning to the conspiratorial ambiance of earlier times could change that.

All of which bemused William Gibson. He was credited with inspiring the idea of virtual reality with his concept of cyberspace. Leary, in one of his rocket-boosted flights into hyperbole, described Gibson as 'Most valuable performing philosopher', the writer of 'the underlying myth, the core legend, of the next stage of human evolution'.[18] Gibson, however, was unsure that cyber-space had been adopted by the virtual reality industry quite in the spirit he originally offered it. 'I sometimes get the feeling that technical people who like my work miss several layers of irony', he said at the Cyberthon.[19]

In an editorial on the 1991 SIGGRAPH conference at Las Vegas, one which featured an entire gallery devoted to VR installations entitled 'Tomorrow's realities', the editor of the trade magazine *Computer Graphics World*, Phil LoPiccolo, wrote that virtual reality was threatening to split the computer graphics world in twain: 'At one extreme are those who predict that virtual reality may some day rank as one of the greatest inventions since the telephone or the personal computer. At the other are those who believe it will be one of the biggest boondoggles since the videophone or the personal helicopter.'[20] Who, then, was right? Phil LoPiccolo was not at all sure. No one was at all sure. Virtual reality had become so entwined with the metaphors and rhetoric that it was losing all definition. It had become a cloud of ideas whose shape depended on who was looking at it: almost in the shape of a technological breakthrough, more like a metaphor, perhaps a tool of political change.

So is there any way of discovering the subject of one of the most important meetings ever held by human beings? Is there any way of discovering the true shape of the cloud? There is, because underneath the rhetoric and the metaphors lie the core issues that aroused the excitement about virtual reality in the first place, that made it both a response to technological and social changes, and which make its emergence so significant. I can predict with absolute certainty that there will not come a day when virtual reality in the sense meant by its most ardent supporters will be realized. Furthermore, most of the research done under its auspices could as easily have been done, and in some cases is being done, under the auspices of more mundane disciplines: computer interface design, computer simulation, 'human factors' (which deals with the human factors that come into the design of technology; the NASA virtual reality research work was done in the agency's human factors division). That is not the point. The significance of virtual reality – and it is one hard to overstate (though Leary, no doubt, could manage it) – is that it directly confronts the question: what is reality?

Notes

1   Judith K. Larsen and Everett M. Rogers, *Silicon Valley Fever*, London: Unwin, 1986, p. 261.
2   Andrew Pollack, 'What is artificial reality? Wear a computer and see', *New York Times*, 10 April, 1989, p. 1.
3   Jay Stevens, *Storming Heaven: LSD and the American Dream*, London: Grafton, 1989, p. 180.
4   Timothy Leary, *Flashbacks*, Los Angeles: Tarcher, 1983, p. 40.
5   Stevens, 1989, p. 263.
6   *The Media Laboratory 5th anniversary*, Cambridge, Mass.: MIT 1990.
7   Stewart Brand, *The Media Lab: Inventing the Future at MIT*, London: Penguin, 1989, p. 113.
8   Stevens, 1989, p. 343.
9   Steven Levy, *Hackers*, New York: Bantam, 1984, p. 172.
10 Ted Nelson, *Computer Lib*, Redmond: Tempus, 1987.
11 Larsen and Rogers, 1986, p. 3.
12 Levy, 1984, p. 40.
13 Sherry Turkle, *The Second Self: computers and the human spirit*, Simon & Schuster, 1985, p. 196.
14 Levy, 1984, p. 415.
15 Benjamin Woolley, 'Love is for wimps and hackers', *Guardian*, 16 November, 1989.
16 Brand, 1989, p. 42.
17 Teresa Carpenter, 'Slouching toward cyberspace', *Village Voice*, 12 March 1991, p. 38.
18 Timothy Leary, 'Quark of the decade?', *Mondo 2000*, Fall 7, p. 52.
19 Carpenter, 12 March 1991, p. 37.
20 Phil LoPiccolo, 'Virtual rewards', *Computer Graphics World*, 14 (9) September 1991, p. 6.

2
SIMULATION

Trying to trace the origins of the idea of virtual reality is like trying to trace the source of a river. It is produced by the accumulated flow of many streams of ideas, fed by many springs of inspiration. The computing industry has, however, had difficulties coming to terms with its offspring having such indefinite origins. Its creation myth is filled with the rhetoric of invention and discovery, of 'founding fathers' and 'pioneers'. Technologists, being mostly American, are fond of titles that evoke their New World heritage. At graphics conferences, the grey-bearded delegates advertise their honorary status with ribbons proclaiming parenthood of the industry they helped create. Awards are given for 'outstanding contributions' and 'achievement'. Like the arts, the sciences cherish the heroes they create.

However, it goes further than that. Technology is often presented as initiating change, some separate activity that every so often introduces the products of its efforts into the world for good or evil. The problem with such an attitude – held, it should be noted, by those who fear technology as much by those who embrace it – is that it has nothing to say about origins. Technology becomes nothing more than the product of genius ignited by that ineffable spark of inspiration, which is a fat lot of good if you want to make some sense of it and the world it is a part of.

This attitude does have its merits. It is romantic, it is simple, it is human, and conveniently marks the limits of any historical account. You trace back to one prime mover, one isolated human act. But what you gain in elegance you lose in richness; you may hear snatches of the melody, but never the symphony. You also get a very poor basis on which to predict the effect the technology will have. The technological myth makers assume that the purpose or function of a new technology is somehow implicit in its origins, the product of genius rather than environment, all nature and no nurture, which is something that was certainly never true of the computer.

Already the hunt is on for the moment of virtual reality's invention. The favourite candidate is the publication in 1968 of an academic paper by Ivan Sutherland entitled 'A head-mounted three dimensional display'.[1] This paper certainly specifies one of the key technologies to be used in subsequent virtual reality experiments, but to understand what Sutherland was aiming at can only be done by working backwards rather than forwards, beginning with an earlier paper, portentously entitled 'The Ultimate Display'.[2]

Sutherland is one of the most influential figures in the history of computing. He has been hailed the father of computer graphics by those who work in the field, as well as a key figure in the commercial development of computer simulation, being a founder of Evans & Sutherland, a leading developer of military aircraft and vehicle simulators. He has enhanced his reputation by refusing to talk to the press about himself or his work, and has generally been held in awe by the industry that regards him to be one of its creators. All of this makes him an excellent candidate for the role of inventor and hero of virtual reality.

'The ultimate display', wrote Sutherland, 'would . . . be a room within which the computer can control the existence of matter. A chair displayed in such a room would be good enough to sit in. Handcuffs displayed in such a room would be confining, and a bullet displayed in such a room would be fatal. With appropriate programming such a display could literally be the Wonderland in which Alice walked.'[3] It was a startling vision, and not just because of the macabre examples. These were, it must be recalled, early days in the history of computing. Computers were still huge, hot, heavy technology, expensive, and capable of producing only the crudest pictures. And here we have a computer technologist anticipating the realization of Wonderland.

How, then, would this ultimate display work? Sutherland was writing at a time when displays were understood to be any kind of device that would show the computer's 'output'. 'Input' and 'output' are key computing concepts. A computer is a device that takes information in – the input, processes it, and then puts out the result – the output. To take a simple example, the process may be to add two numbers together. The input information would be the two numbers to be added, say '1' and '2'. The output information would be the sum of those two numbers added together, '3'.

Displaying the output is not as straightforward as it sounds. The computer simply produces a series of electrical impulses that within the system represent the number 3. The display device then has to turn this into something intelligible. It could do this by producing three beeps on a loud speaker (computer jargon allows such oddities as audio displays). More likely, it will turn the electrical impulses into a varying current that manipulates an electron beam that inscribes the shape of a three on a screen – a visual display unit, VDU.

Sutherland's ultimate display, however, would go much further than that. It would not be just visual, or even just aural, it would be 'kinesthetic'. 'The force required to move a joystick,' wrote Sutherland, 'could be computer controlled, just as the actuation force on the controls of a Link Trainer are changed to give the feel of a real airplane.'[4] Here we encounter the reason why the origin of any scientific or technological idea is harder to discover than is often assumed. Having arrived at what seemed to be the source of a virtual reality, we discover that it is being fed by another stream of ideas, that of flight simulation.

Fﾘﾘﾆﾊﾟ Sﾉﾏﾘﾘﾑﾟﾘﾟ

* * *

The danger of flying planes was obvious almost from the moment of their invention. The world's first major air accident, which seriously injured Orville Wright and killed his co-pilot, happened just five years after the Wright brothers' first (if, indeed, it was a first; that is contested by the French) flight at Kitty Hawk. The occasion of the accident was a flight trial laid on for the American War Department. The mix of military potential and inherent danger made the development of the simulator, which could train pilots quickly and cheaply as well as safely, inevitable. Designs were being patented as early as 1910.[5]

The Link Trainer referred to by Sutherland was developed by Edwin Link in the late 1920s. He was born into the then highly successful and technologically adventurous world of mechanical musical instruments. His father had founded the Link Piano and Organ Company of Binghampton, New York, and Edward was awarded his first patent for a technique that improved the mechanism of player pianos. The player piano – also known as the 'Pianola', after the brandname of another piano-player manufacturer, the Aeolian company – was an impressive demonstration of the power of automation. By using a series of pneumatic switches to play the piano's keys, and activating the switches by means of a series of holes punched in a ribbon of paper, it was possible to reproduce an entire musical performance. The piano seemed to 'come alive'.

The success of the technology at reproducing the dynamics of piano playing inspired Link to see if pneumatics could be used to provide a way of reproducing aircraft movements. He patented his first 'trainer' in 1930, describing it as 'a novel, profitable amusement device' as well as 'an efficient aeronautical training aid'. Just over 50 years later, one of the world's first commercial virtual reality systems, developed by a British company called W Industries, was launched using more elaborate language ('the only limit is your imagination'), but promising strikingly similar attractions: an amusement device that would profit arcade operators by providing players with the sensation of flying an aircraft.

Link's first trainer provided no instrumentation, simply a cockpit mounted on a fixed base that 'pitched', 'rolled' and 'yawed' in response to the movement of a primitive set of controls operated by the 'pilot'. A charming feature of the machine was that it looked like a baby-sized plane, complete with wings and tailplane. The official function of these was not, obviously, to make the machine airworthy, but to provide both the student pilot and training instructor with a recognizable indication of the position of the controls. Gradually, such features disappeared. By the late 1930s the wings were more like those of an ostrich, barely visible, almost a parody of the real thing. Over the years, flight simulators have become physically less and less like aircraft, losing any visual reference to aircraft fuselages, leaving behind a windowless dome mounted on a hydraulic platform, an airline insignia the last remaining external evidence of what the machine is supposed to be simulating.

The reason for this gradual abandonment of redundant features is obvious: there is no need for fake wings if everything else – movement, instrumentation, controls – is realistically reproduced. But it is harder to understand why they were there in the first place. As far back as 1917, designs for trainers included propellers and aerodynamic bodywork. Did the designers really think that these would fool the pilots in believing that they were in a real aircraft? Obviously not. The reason has more to do with the fact that early trainers were not simulators at all. They were more like fairground rides, they did not attempt to simulate aircraft so much as suggest a vague empirical likeness. Had they been given portholes and propellers fixed at the rear instead of the front, they could have easily been used as submarine trainers.

Link trainers, however, did manage to reproduce some of the feel of flying. The pneumatic mechanisms were advanced enough to mimic an approximation to the sorts of movements a pilot would experience in a real cockpit. They could also reproduce the forces that the pilot feels through joystick in real flight – these being the 'actuation forces' that Sutherland later imagined being exploited in his 'ultimate display'. But the trainer could only capture at the crudest level the relationship between joystick movement, the plane's disposition and the position of the flaps. It could not accurately represent the way a plane actually flew, the relationship that exists between changes in the controls and the effect on the aircraft's position, speed and altitude. It could imitate flying, but it could not *simulate* flying. To simulate something you need more than mere mimicry, more than an ability to produce actions that are like the ones you are wanting to simulate. You need a working model.

The distinction between simulation and imitation is a difficult and not altogether clear one. Nevertheless, it is vitally important. It lies at the heart of virtual reality.

* * *

An altimeter, an instrument for measuring altitude, is a standard piece of aircraft equipment. It indicates the height at which the plane is flying by measuring the atmospheric pressure outside the aircraft. Atmospheric pressure falls as altitude increases. The relation of one to the other follows a very simple formula: for every 16 kilometres (10 miles) of altitude, pressure drops by a factor of about ten. An altimeter, therefore, simply needs some connection that links its gauge to a pressure meter, such that as the pressure meter's reading falls, the altimeter's needle rises by an appropriate ratio.

If this is done mechanically by means of gears, the resulting linkage is a mechanical model of the altitude/pressure relationship. But the significance of this relationship can be taken to a further level of abstraction, since it can be expressed as a simple mathematical formula. The actual arithmetic used will be based on 'differential calculus', the type of maths used to deal with related quantities and rates of change. This is because it is the rate of change in pressure as the aircraft rises (or falls) that is used to calculate the altitude. If the pressure drops by one-tenth since it was last measured, we know that we are just over one and a half kilometres higher than we were before.

Calculus is a form of maths few of us ever encounter, because it is usually just outside the standard maths curriculum. However, it uses nothing that is not part of conventional arithmetic and therefore can be used to specify an 'algorithm', a procedure that will yield the result given the necessary 'inputs', or starting values (in this case, the change in pressure). The term algorithm is named in honour of Mohammed ibn Musa al-Khuwarizmi, a ninth-century Arab mathematician whose second book gave us the word algebra (from the Arabic *Al-jabr*). It is now used to describe any mathematical procedure that can be completed by some automatic, mechanistic means, without the need for any human imagination or creativity. Not all mathematics, as mathematicians sensitive to the charge of being unimaginative point out, is algorithmic.

Another example of an algorithm might be a formula to work out the change in a bank balance over a given period produced by a given interest rate. If the interest rate is, say, 10 per cent a year, that means that given a current balance of perhaps £100, you can work out, without any further information nor so much as a glimmer of inspiration, that in a year's time the balance will be £110. Furthermore, formulae can be used to generalize from the specific instance to a general rule that can be used to calculate the balance produced by a given interest rate applied over any given period of time. The only 'inputs' that must be given to produce an 'output', a result, is the current balance, the interest rate and the length of the period. Such a formula can be quite complex. For example, if, as is the case with some accounts, the interest rate itself rises as the balance rises, the interest accrued can push up the balance sufficiently to increase the interest rate, producing a sort of feedback loop (or, if the balance is in the red, a vicious circle). Even this sort of equation can still be reduced to an algorithm. It may, in other words, be a more complicated procedure, but it will remain a mechanical procedure nonetheless. However, it is a procedure that must be worked out at the rate at which interest is being earned. It is a 'real time' problem, since the period between interest payments will influence the amount of interest paid. With real time systems, it is not just what the algorithm does that matters, but when it does it.

Inspiration or creativity can help work out an algorithm – someone might be able to guess the result accurately – but neither is necessary in order to reach a solution. Perhaps it is because formal maths education is about algorithms, rather than the more inspirational aspects of mathematics, that many people (myself included) form such an unsuccessful relationship with the subject. Applied mathematics, which is promoted as being more exciting because it is more 'relevant', could be dull because it turns us into machines.

Conversely, because algorithms are mechanical, they can be worked out by machines. The machine only requires mechanisms that can perform basic arithmetical functions such as addition and subtraction. Such mechanisms are all around us. A lever or a set of gears, for example, can be used for multiplication, the degree of multiplication varying according to the position of the fulcrum or the relative size of the cogs. Charles Babbage, the Victorian engineer and one of the many people to be credited with the invention of the computer, designed a 'Difference Engine' to perform the repetitive calculations that clerks (known as 'computers' before the word was applied to machines) had to make to draw up the tables that were becoming increasingly important to industry and trade. The configuration of the cogs was complex and ingenious – for example, Babbage developed an elaborate mechanism for 'carrying' when the sum of two numbers was greater than ten. But the cogs were just cogs; there was no magic to the mathematical procedures that resulted from their grinding.

In the 1930, Vannevar Bush found that he was having problems working out the equations needed to predict the ebb and flow of electrical current produced by power companies and distributed by the increasingly tangled and tenuous grids that were spreading across the US. What would happen if there was a sudden increase in demand for electricity of, say, 10 per cent? Would it overload the system and cause a power cut? What if the increase was 100 per cent? Would the problem simply be ten times worse than if it were 10 per cent? Such questions could not be answered using one, eternally correct calculation, just as a film cannot be adequately viewed by inspecting one of its frames. They could only be solved by differential calculus, equations that themselves do not yield one result, but another equation, which is then used to find a result for each particular circumstance.

Bush's machine for working out such equations was a room-sized deck of rods and cogs he called a 'differential analyzer'. This provided a means of analysing the results produced by a particular differential equation, the series of starting values being represented by a graph which operators entered into the machine by tracing its shape with a pointer, and the result being represented by another graph traced by the machine on a plotting table.

The mechanization of differential equations was the break-through that made possible the mathematical modelling of moving, changing phenomena like the flight of an aeroplane. Flying is partly about numbers, the numbers represented on the plane's instrument panel: the degrees on the compass, the knots on the airspeed indicator, the feet on the altimeter. These numbers vary according to the way the pilot manipulates the controls: the joystick, the pedals. These manipulations can also be turned into numbers: the joystick can, for example, be connected to two variable resistors, the sorts of electrical devices used for volume controls in hifi systems which vary the voltage passing through them as they are turned up and down. If one of these variable resistors is connected so that it is turned as the joystick is moved forwards and backwards, and another is connected so it is turned as the joystick is moved from side to side, two varying voltages are produced which vary as the joystick is moved. Connected to a pair of voltmeters, these voltages can be represented as numbers.

The question that flight simulator builders like Link began to ask themselves during and after the Second World War was whether there was a way of working out the equations that produced the first set of numbers, the readout from the instruments, from the second, the numbers on the voltmeters representing the position of the controls. Vannevar Bush's differential analyser at least showed that it was possible to use a machine to solve complex differential equations; the problem was finding a technology that could solve them in real time, at a rate that reflects the rate at which the position of the controls is changed.

By the late 1930s, however, no such technology existed. Instead, flight trainer designers had concentrated their attention on developing better means of mimicking the experience of flight, for example, by reproducing what the pilot might see out of the cockpit window. In 1939 Link designed a 'celestial navigation trainer' for training bombers to navigate by the stars during long-haul, night-time missions.[6] The result was a monstrous piece of fairground equipment, 45 feet high, with a cabin large enough to accommodate a crew of three, a movable dome studded with lights representing the stars and a screen beneath the bomb aimer's station to show the targets over which the bomber would fly.

Even as late as 1943, player-piano technology was still forming the basis of new trainer designs. The 'Silloth Trainer', for example, was designed by an engineer working for Link's competitor, Aeolian, and used the same combination of pneumatic pumps and valves to reproduce aircraft motion. It did, however, introduce two new and important developments: the reproduction of a whole ensemble of in-flight phenomena, even cockpit noise (generated by a Hammond organ), and the ability to reproduce the flying characteristics of different types of aeroplane.

However, by now player-piano engineering was approaching its coda. The Second World War was forcing the development of the technology that was to make true simulation possible: the real-time, digital, electronic computer, the machine that could process any algorithm, and do so incomparably faster than any mechanical device.

War is the blast furnace of technological innovation, melting old ideas down and allowing new ones to form. At one level, the First World War was about non-communication. Deafened by shellbursts, blinded by gas and commanded by generals remote from the battlefield slaughter, the soldiers' experience seems to have been one of almost total sensual isolation. Yet this was the war that laid the foundations of mass communications. Before the war's outbreak, the verb 'to broadcast' was still an agricultural term, used to refer to the scattering of seed. Similarly, the Second World War marked the period when the term computer changed from meaning a clerk who performed calculations to meaning a machine that could be programmed to perform different types of calculation.

The impetus to develop such a machine was mainly military, specifically the need for ballistic tables for predicting the trajectory of bombs and missiles. Before computers, tables were vital to any sort of attempt to predict the behaviour of any sort of dynamic, changing system. If a gunner wanted to work out whether or not a long-range projectile would meet its target, he would have to find the appropriate entry in a ballistic table, one that approximated best to the missile's weight, the force to be applied to it and the atmospheric conditions through which it would travel. Innumerable methods were explored to find a way of making the compilation of such tables faster and more accurate. The behaviourist B. F. Skinner even suggested a way of dispensing with the need for them altogether. He proposed packaging trained pigeons into a 'homing' bomb, and using them to help guide the projectile to its target.[7] The idea never got off the ground.

The development of Electronic Numerical Integrator and Computer, ENIAC, at the University of Pennsylvania fared better. This huge machine was designed to take some of the pressure off the US Army's Proving Ground at Aberdeen, Maryland, where, by 1944, requests for new ballistic tables were coming in at a rate of six a day. A single, 60-second trajectory could be worked out by hand in about 20 hours, on a Bush-type differential analyser it would take about 15 minutes. The army, however, wanted to calculate at a rate appropriate to the era of high-yield kill rates and mass-destruction lines. It needed figures for not just one trajectory, but a whole range, and for a whole range of different conditions.

And so it commissioned the ENIAC, which promised to calculate each trajectory in 30 seconds.[8]

As the designers of ENIAC pointed out, however, their machine could do more than compile ballistic tables. 'There is no essential or fundamental restriction imposed by the ENIAC design on the character or complication of the problems which it can do', they wrote. 'The ENIAC should be able to solve many of these larger problems faster than they can be done on any other existing machine. Undoubtedly, the ENIAC will be used in this way . . . until some other machine with better facilities for handling non-linear partial differential equations can be built.'[9]

Over the following few years there was to be many other machines, quickly establishing, among other things, the computing industry's taste for acronyms: EDVAC (Electronic Discrete Variable Automatic Computer), EDSAC (Electronic Delay Storage Automatic Calculator), UNIVAC, SEAC, SWAC and even MANIAC. Each of these tried out new technologies, but all of them had the same basic design, the same architecture'. They were electronic, which means that operations were performed by changes in electrical currents passing through circuits rather than the movement of physical mechanisms. They were digital, which means that they dealt with discrete quantities – numbers, in other words – rather than analogue, like the differential analysers. These two features meant they were both faster and more accurate (though some contested the accuracy claim) than the mechanical calculators they replaced – fast and accurate enough, indeed, to work out a missile's flight path in real time, or even faster as the ENIAC had demonstrated by calculating a 60-second trajectory in less than 60 seconds.

In 1944, around the same time as the ENIAC was being built, a group under Jay W. Forrester at the Massachusetts Institute of Technology's Servomechanisms Lab began work on 'airplane stability control analyzer' for the US Navy. However, the project had a higher ambition, to use this new, electronic, digital technology to create a generalized flight trainer, a machine that could compute the equations of aircraft motion and aerodynamics. 'Putting wind tunnel data into the trainer would cause it to fly like an airplane not yet built', promised Robert R. Everett, a member of the project team, perhaps a little optimistically. Project Whirlwind, as it came to be called, was completed for testing in

1949 and was fully operational by 1951, just in time, as it turned out, for the US Navy's funding to run out as the war had come to an end, and for new funding to be found from the Air Force, which wanted to use the system for air defence.

However, while the Whirlwind was still being tested, the project researchers began to explore characteristics that had never been intended as part of the original design. In late 1948 and early 1949, they began to play around with the oscilloscope screens that were used to display system information. They noticed that, by issuing appropriate instructions to the computer, the screens could be manipulated to create patterns. They even managed to create a game using a dot that traced the path of a bouncing ball, the game element coming from adjusting the input variables (in other words, the height from which the 'ball' was dropped) so that it would fall through a 'hole' in the 'floor'. Though no Pacman, it graphically demonstrated how a computer could be used to reproduce the animated, moving 'dynamics' of a real ball game using nothing more than the pure maths, the manipulation of abstract symbols that have nothing in particular to do with balls or the way they happen to bounce. The result was claimed to be the first example of 'man machine interactive control of the display', as well as the first computer game.[10]

An early testing program written to run on the Whirlwind produced a series of dots arranged as a grid on the oscilloscope. These dots represented the states of the system's memory devices ('storage tubes', as they were called). The researchers wanted a way to discover which dot represented which particular tube. To do this, they built a device they called a 'light gun' (this was, after all, just after the war), comprising a tube with a light-sensitive detector at one end. By connecting this to the Whirlwind, and pointing it at the oscilloscope's screen, it was possible to 'select' one of the dots (the computer could identify which one it was by checking through each dot in turn and logging the one that triggered the light gun detector) and thereby identify its associated storage tube.

As far as technological breakthroughs go, neither the game nor the light gun were particularly revolutionary. But they demonstrated two of the more important possibilities opened up by fast computation: the use of pure mathematics to simulate real dynamics, and the possibilities of enabling humans to 'interact' with such simulations. These two elements are, of course, the basis of computer simulation and virtual reality.

The discovery of the principles of computer simulation by those researchers must be seen as of equal importance to the discovery of movies. The Whirlwind researchers had managed to get the dot to behave as a ball does – to all intents and purposes, to *be* a ball. Such a transition from imitation to simulation was to prove as important to computing as the discovery that images projected in quick succession could reproduce motion was to prove to the development of movies. We forget how unlikely an illusion films must have seemed to those who first saw them flicker. The matter-of-fact use of the word 'animation' disguises the word's original, mystical association with the divine gift of bestowing life. The Whirlwind had achieved an equivalent breakthrough by animating a ball that had no physical existence – or even imaginary existence, since no human imagination was needed to conceive it.

At around the time Whirlwind was first deployed, the developers of ENIAC at the University of Pennsylvania were also using US Navy funding to develop a real-time simulation system, the Universal Digital Operational Flight Trainer, UDOFT. Unlike Whirlwind, this was to reach its original aim in 1960 of demonstrating the feasibility of building a computer that could model aerodynamics. By the early 1960s, Link had developed its own flight simulation computer, the Mark I, helping the technology to move from the academic and military world into the commercial and civil one. The Mark I was a machine that simulated an aeroplane's flight dynamics rather than merely mimicked them. With the pilot's manipulation of the controls acting as input, computers such as the Mark I could use equations modelling the characteristics of a particular plane to output the whole multi-sensory ensemble of information the pilot would experience: the landscape seen through the cockpit window, wind and engine noise, the readings on the aircraft's instrumentation, and the feedback forces felt through the joystick and pedals. The result was a highly specialized version of the 'ultimate' display that Sutherland described in 1965.

* * *

Sutherland's great imaginative leap in his 1965 paper was to generalize the idea of simulation. The cockpit provided a highly controlled 'interface' between the pilot and the simulation of a flying aircraft. Would it, he speculated, be possible to provide an interface that worked with any simulation, one that reproduced not just the experience of flying in an aircraft but of being in any artificial space, a space 'within which the computer can control the existence of matter'? It would be unfair on his technological predecessors to credit him with the 'invention' of this idea, many had before him hinted at it in different ways, but he was one of the first people to give it clear expression, to find the underlying 'meaning' of simulation, though it was, as we shall see, the developers of virtual reality who were to give the idea its most forceful expression.

The more prosaic title of Sutherland's 1968 paper, 'A head-mounted three dimensional display' reflected its practical emphasis. It outlined the results of work undertaken at Harvard and funded by the same defence sources that had financed the development of the Whirlwind and ENIAC computers. The aim was to 'present the user with a perspective image which changes as he moves'.[11] Sutherland's display comprised two main elements, a helmet and a tracking sensor. The helmet made its wearer look like an exotic beetle, with two grotesque insect antennae emerging from each eye. These were the wires and bodywork of two tiny TV screens, one for each eye, which filled the user's field of view with a computer generated picture. The tracking sensor monitored the user's position and movements. Sutherland experimented with two types: one uncomfortably attaching the cranium of the helmet to the ceiling via an adjustable rod, the other using an ultrasound source fixed to the helmet that was located by receivers set above the user's head.

The image the user saw through this strange apparatus was one generated by pure maths. Using a coordinate system, it is possible to describe a three-dimensional space mathematically. Imagine a sheet of graph paper, with numbers up one side and along the bottom identifying each row and column of the paper. The square at the bottom, left-hand corner could be described by two numbers, '0,0', the square at the top right-hand corner as, say, '100,100' (assuming the graph paper has 100 rows and 100 columns) – this is our two-dimensional artificial space. You could

describe a square in the space by identifying each of its corners
with a pair of numbers – its coordinates – the first representing
the column number, the second the row number: 10,10; 90,10;
10,90; 90,90. Imagine, further, that instead of using pairs of
numbers to identify each corner, three numbers were used, the
first two representing, respectively horizontal and vertical position,
the third representing the third dimension. The coordinate
10,10,10 will then represent a point in three-dimensional space
ten units along, ten units up, and ten units back (or whichever
direction the third dimension is in).

Using a computer to describe three-dimensional shapes in this
fashion is routine. Furthermore, it provides a means of manipulating them, too. For example, the two-dimensional square described
by the coordinates 10,10; 90,10; 90,10; 90,90 could be made bigger
simply by increasing the numbers representing its extremities:
10,10; 100,10; 10,100; 100,100. Similarly, the square could be
'moved' by adding to or subtracting from all the coordinate
numbers. The same could be done using the coordinates representing a cube, with the numbers added to or subtracted from the
coordinates representing the cube's corners providing a means of
'moving' the cube around and changing its size. In the computer's
memory, of course, the cube is just a set of numbers. But with the
appropriate display, a screen or printer, these numbers can be
turned into a picture of the cube.

The problem with screens and printers, however, is that they
produce a two-dimensional image. Sutherland's idea was to use
the head-mounted screens to simulate what it would be like to
'look around' models of three-dimensional objects. The physical
space monitored by the tracking sensor is mapped by the
computer onto the mathematical space of the objects. As the
helmet's position and orientation changes, the computer works
out from the tracking information how the objects will appear if
seen from the equivalent position in the mathematical space – the
objects are made bigger and smaller as the the user approaches
and retreats; their perspective changes with the angle from which
they are viewed. As a result, the third dimension is reconstituted
in what the user sees, just as movement is reconstituted by
showing a series of still images in quick succession.

Of course, few shapes are as simple to model using a computer
as a cube. Any natural form – a cloud, a twig, a human – is

extremely complex to describe, because it does not have the pure regularity of geometric shapes. Also, there is more to the appearance of an object than its shape: there is the texture of its surface, and the position and nature of the light illuminating it (there is no sunshine inside a mathematical model). But Sutherland was not all that concerned with practical limitations. He had established the principle that any shape that could be described mathematically could exist in this computer-generated space, and that a head-mounted display of the sort he proposed provided a chance to peer into this space and inspect whatever was within it.

With the 'ultimate display', the objects in computer-generated space would not just be visible; they would be tangible. Being mathematical, physical laws can be applied to reproduce qualities like mass just as geometrical ones reproduce size and shape. Using the ultimate display, such 'physical' objects could be picked up, dropped, thrown, even moulded. The force actuators in a Link-like joystick could, Sutherland imagined, be used to simulate the sensation of 'pushing' an object seen through the head set, the effort required being proportional to the object's 'weight'.

What, then, is the status of such simulated objects? Are they the product of fiction, of imagination? Not in the usual sense. Are they, then, natural, physical, actual? Again, not in the usual sense. They exist in what Sutherland called a 'mathematical wonderland', a wonderland ultimately determined not by the imagination of any particular author, but by the laws of mathematics. One could imagine that a designer might use a computer-aided design program to create a series of objects for inspection using a head-mounted display, and attribute the design of those shapes to the designer. But what is important about them from Sutherland's point of view is that they are mathematical objects – this is why they have the strange form of independent existence revealed by his ultimate display. The helmet reveals their complete shape, the kinesthetic display their physical characteristics – chairs become 'good enough to sit in', bullets real enough to kill. There can be no 'author' of such objects in the same sense that there is an author of a book (though, as we shall see, even the literary author, according to some French theorists, no longer exists). They partake in a new mode of existence, a mode that is not actual nor yet imaginary – the mode that has been called virtual.

Notes

1 Ivan Sutherland, 'A head-mounted three dimensional display', *Proceedings of the Joint Computer Conference*, 1968, 33, pp. 757–64.
2 Ivan Sutherland, 'The ultimate display', *Proceedings of the International Federation of Information Processing Congress*, 1965, pp. 506–8.
3 Ibid., p. 508.
4 Ibid., p. 507.
5 J. M. Rolfe and K. J. Staples (eds) *Flight Simulation*, Cambridge: Cambridge University Press, 1986.
6 Rolfe and Staples, 1986.
7 Charles and Ray Eames, *A Computer Perspective*, Cambridge, Mass.: Harvard University Press, 1990, p. 131.
8 Ibid., p. 133.
9 J. P. Eckert Jnr, J. W. Mauchly et al., *Description of the ENIAC and Comments on Electronic Digital Computing Machines*, Moore School of Electrical Engineering, University of Pennsylvania, 1945.
10 Norman H. Taylor, 'Retrospectives: the early years in computer graphics at MIT, Lincoln Lab and Harvard', *Computer Graphics*, 23 (5), December 1989, p. 21.
11 Sutherland, 1968, p. 757.

3
VIRTUALITY

'IBM launched a product/concept called "virtual memory" way back in the '70s with panoramic ads showing the silhouette of a city skyline with the caption "a new day is dawning"', wrote a one-time salesman for the electronics company Ferranti in 1991. 'This irritated me because 11 years previously Ferranti had brought out the same idea of swapping pages of memory out of store and disk and had called it "paging."'[1] If it was my choice, I know which word I would prefer. 'Paging' is what bellboys do in hotel foyers. 'Virtual' was and remains a much grander word, scandalously underused, a huge vessel of semantic vacuity waiting to have meaning poured into it. Computing has provided some of that meaning, initially a quite modest meaning, but increasingly a cargo worthy of its carrier.

IBM introduced 'virtual memory' with two new mainframe computers, the 370/158 and 370/168 systems, introduced in August 1972. By the stage of their launch, IBM had established itself as the dominant supplier of the world's computing power. The 360 range of machines, introduced in 1964, had become a symbol of American corporatism, the foundation of centralized, mechanized, depersonalized multinational companies and state utilities, the purveyors of payrolls and bills. The '360' designation was, according to evidence submitted to one of the many anti-monopoly suites brought against IBM, a reference to the number of degrees on the compass, reflecting the machine's all-encompassing nature – which, at least in terms of global market penetration, it certainly achieved.[2] The 370 range would, by the same logic, suggest an all-encompassing ambition beyond even IBM. In a sense, however, 'virtual memory' was designed to perform just such a trick. The idea, as the disgruntled Ferranti salesman suggested, was not new; it had been explored as a method of overcoming one of the main constraints of computer design.

The constraint was memory. 'Memory' is the rather grand, perhaps misleadingly anthropomorphic, term for the technology used to store information for computer processing. There are two sorts of memory, sometimes simply known as fast and slow. Fast memory is the sort accessed by the computer's processor, the electronics that perform the computations, directly. Slow memory is for keeping information that is not needed immediately and for which there is not enough room in fast memory. Fast memory tends to be electronic, once an array of magnetic 'cores' (some diehard computer users still call it 'core memory'), now a tiny silicon chip. Slow memory tends to be electro-mechanic, taking the form of disks and tape.

At the time IBM was developing its 370 range, fast memory suffered from two main constraints: it was expensive and bulky. This limited its capacity and, in turn, limited the size of computer programs and amount of data the computer could process at any one time. The aim of virtual memory was to create more room in fast memory than was actually provided by the computer's physical memory circuits. The system would then make up the difference by invisibly swapping segments or 'pages' (hence the alternative term 'paging') of information from slow into fast memory as and when they are needed. It is a system familiar in other walks of life. A shop, for example, does not have its entire stock behind the counter, but 'swaps' goods in and out of the stockroom in accordance to customer demand. Of course, such a system has its limitations: there will be unanticipated requests for slower-moving goods not usually stored on the shop's shelves, forcing the shop keeper either to disappoint the customer or rummage around in the stockroom. Similarly a program running in virtual memory will occasionally be caught by an unexpected request for some information that has not been moved into fast memory, forcing a delay in the system while the program attempts to collect the appropriate page of information from disk. But at least such a system means the shopkeeper can sell more goods than will physically fit on the shop's shelves, and the computer program can manipulate more information than will fit into the system's fast memory. The term 'virtual' cleverly captured the status of paged memory. The fact that virtual memory is not real memory is a mere technicality; from the point of view of the computer and its user, virtual memory is just as real as actual memory. It is a simulation of a physical system that is perfect in every detail, except that it might be slower than 'real' memory.

'Virtual' has a respectable pedigree as a technical term, going right back to the origins of modern science. It was used in optics at the beginning of the eighteenth century to describe the refracted or reflected image of an object. By the beginning of the nineteenth century, physicists were writing of a particle's 'virtual velocity' and 'virtual moment'. The word is still used in physics to describe the exotic behaviour of subatomic particles that appear so fleetingly they cannot be detected. It has come a long way from its original use as the adjectival form of 'virtue', in the days when virtue itself meant to have the power of God. Echoes of that early meaning, however, survive in the excitable claims of virtual realists to have the power to create their own worlds. And it is appropriate that the word should resonate with a certain amount of divine significance, because the computing concept of 'virtual' is much more than a matter of mere technology. It means something that goes to the scientific heart of reality.

Every computer is virtual, each one a shadow of one machine, a machine specified, though not built, in 1936 by the British mathematician Alan Turing in a paper entitled 'On computable numbers'.[3] The temptation is to say that Turing's paper was a turning point in the intellectual history of our times, but it was published in a period when every point was being turned. A reliable, scientific basis of knowledge was simultaneously being constructed and destroyed. Turing's paper was part of this process, helping to reinforce breakthroughs that had left much broken.

The foundations of mathematics seemed pretty secure after Isaac Newton had demonstrated their ability to describe the machinery of the universe. They needed no underpinning when, at the beginning of the nineteenth century, the French mathematician Pierre Laplace proclaimed that, given adequate knowledge about the state of 'all the forces by which nature is animated and the respective situations of the beings who compose it – an intelligence sufficiently vast to submit these data to analysis – it would embrace in the same formula the movements of the greatest bodies and those of the lightest atom; for it, nothing would be uncertain and the future, as the past, would be present to its eyes'.[4] Poor old Pierre has since had to take the rap for excessive scientific hubris – 'Laplacian', it is called. No one now believes, of course, that such a perfect knowledge of the universe would be possible. His timing could have been better, too. At around the same time he was proclaiming that the scientific theorists could pack up and leave the job to the measurers and quantifiers, Carl Friedrich Gauss, a mathematician ranked by some alongside Archimedes and Newton,[5] published his *Disquisitions*, which demonstrated a method for constructing shapes that could not be described using Euclid's geometry, the first sign that classical and undisputed laws stretching back to the third century BC might not be the end of the matter.

By the beginning of the twentieth century, mathematics had changed out of all recognition. It was no longer just a branch of science, a means of measuring and counting physical phenomena. It had become an apparently self-contained system for manipulating its own symbols, one that operated quite separately from the real world. As Andrew Hodges, a mathematician and biographer of Alan Turing, put it: 'A rule such as "x + y = y + x" could be regarded as a rule for a game, as in chess, stating how the symbols could be moved around and combined legitimately. The rule might possibly be *interpreted* in terms of numbers, but it would not be necessary nor indeed always appropriate to do so.'[6]

Peeling away mathematics from physical reality was liberating on the one hand – it meant that it was no longer the servant of science but the master of itself – but robbed it at the same time of its authority. If it was no longer a means of describing physical reality, what was it for? What truth did it tell?

* * *

Our language betrays an uncomplicated attitude to space. Space is two-dimensional: we have revolutions, go round in circles, get straight to the point, square up. The most sophisticated geometric shape to make it into everyday language is, thanks to the US military, the pentagon. Spheres of influence are rare examples of a linguistic acknowledgement of the third dimension. In mathematics, however, space is treated very differently. Physical space is merely one three-dimensional (four, if you count time) version of it. Mathematical space is more like a graph, but a multi-dimensional one in which there are as many axes as there are properties to be plotted. Chess, for example, can be thought of as the universe of all legal chess moves, and a particular game a path through that space.

It was David Hilbert who formalized this concept of abstract space. In the process, as though almost by accident, he provided a means of explaining the very unreal phenomena displayed by newly discovered subatomic particles such as electrons. Such particles behave very strangely, at least as mapped in physical space. But, the Hungarian-American mathematician John von Neumann, the man credited with developing the basic design principles of all modern computers, noticed that a far clearer picture of how these particles behaved resulted from putting them, so to speak, in Hilbert space. It was as though they belonged more comfortably there than they did in physical space.

Von Neumann's discovery was an astonishing vindication of the power of abstract mathematics. It seemed to demonstrate that abstract mathematics underlay physical reality, not the other way round. What, then, underlay abstract mathematics? What gave it the right to claim access to the truth?

In 1900, David Hilbert gave a speech to the International Congress of Mathematicians in Paris. Equal to the moment and location, Hilbert challenged his peers to solve the outstanding problems in mathematics – indeed, being the mathematician he was, he even enumerated them, counting 23 in all. The second of these was concerned with the basis of mathematics itself. The Italian logician Guiseppe Peano had worked out what he saw as the foundations of mathematics by formulating them as a series of laws or 'axioms' based on logic. Hilbert asked how anyone could be sure that such axioms did not produce an inconsistency. They had failed to reveal one so far – mathematics was apparently 'clean' – but there was no proof, no guarantee that there might not be some inconsistency which might be operating hidden in the background, or which might emerge in the future as mathematics became yet more complex.

The importance of this problem lay not only in its concern for consistency. It also raised the issue of the nature of mathematics itself. Was it 'axiomatic'? In other words, are there a set of basic principles from which all mathematics is derived? And if so, where do those principles come from? At the turn of the century, the proposed answer to the last question was that the principles came from logic.

Being a branch of philosophy, the purpose of logic is the subject of some debate. At its most confident, it is claimed to be about establishing the structure of knowledge and reasoning. It established, for example, that statements of the form 'if p and q then q' (as in 'if it is true to say that it is hot and sunny, then it is true to say that it is sunny') are always true. Like mathematics, logic is abstract and uses symbols. The idea of using it to provide the basis of mathematics was too tempting to ignore, and at the turn of the century, a number of philosophers, notably Bertrand Russell and Alfred North Whitehead, worked away feverishly to achieve this, in the hope of thereby creating a complete, consistent system of pure knowledge.

The aim of making mathematics 'pure', purging it of the uncertainties of physical reality, intensified in the churning wake of the First World War. In the early 1930s, a group of France's finest young mathematicians set up a secret club to publish a mathematical treatise under the collective nom de plume Nicolas Bourbaki, a name of obscure origins, perhaps in part a black joke about the French general Charles Bourbaki who led a disastrous campaign in the Franco-Prussian War that resulted in the death of 10,000 men. Among the club's founding membership were some important names of twentieth-century mathematics, including one Szolem Mandelbrojt who, as we shall later see, also had close ties with a mathematician who probably represents the polar opposite of Bourbakist purity.

Mourning the loss of a generation of mathematicians killed by the war, impatient with the older professors who were left to instruct them, the Bourbakists decided to embark on a project to produce a definitive text of mathematical knowledge, the *Eléments de Mathématiques*, that would survive any future cataclysm. They no longer wanted their beloved subject to rely for its record on the frailty of human memory. They wanted to provide an indestructible account of mathematical knowledge relying on abstract principles and formal proofs.

The same motives underlay Hilbert's restatement at the 1928 International Congress of Mathematicians of the need for some eternal, undeniable validation. As well as being consistent, maths, he demanded, must be 'complete' and 'decidable': every mathematical statement must be provable or disprovable, it should admit of no 'don't knows', and there must be a definite procedure – an algorithm – for establishing whether or not it is true.

Hilbert only made such demands because he confidently expected that they could be met. 'There is no such thing as an unsolvable problem.'[7] Unfortunately, fate yielded to the temptation. Just three years later, the Czech-born mathematician Kurt Gödel published a paper entitled 'On formally undecidable propositions of *Principia Mathematica* and related systems'. The title said it all, or at least most of it: by 'related systems', Gödel meant all those that Hilbert was calling upon his colleagues to develop, indeed any system that attempted to fulfil Hilbert's criterion of completeness.

Gödel's method for puncturing Hilbert's optimism was itself a brilliant exercise in abstraction. He developed a 'theorem' demonstrating that any usable axiomatic system – the sort of system that was supposed to give mathematics its power over truth and falsehood – would not be able to prove or disprove every statement it could express. As a test case, he began by examining ways of formulating the axioms of arithmetic as a set of numbers, so the axioms themselves became part of the system they underpinned. What he discovered was that such a system would generate statements that referred to themselves, including statements of the form 'this statement is unprovable' (only, of course, the statement would be a set of symbols, a formula, not a sentence), which is false if proved true, and true if proved false. This discovery destroyed at a stroke all hopes of creating a self-sustaining mathematical system. Gödel had shown that, to establish the 'truth' of such a system, there had, at some point, to be an appeal to an outside authority, human, divine, whatever.

* * *

In 1928, Bertrand Russell wrote: 'Machines are worshipped because they are beautiful, and valued because they confer power; they are hated because they are hideous, and loathed because they impose slavery.'[8] In 1936, Charlie Chaplin's *Modern*

Times was released. It was film that captured the ambiguous attitude to the 'machine age' of the prewar years, an attitude that has persisted well into the latter part of the century. Chaplin both submitted to technology – it was his first talkie – and attacked its dehumanizing influence. That same year saw the publication of 'On computable numbers'. In it, Turing disposed of the last of Hilbert's original criterion for establishing the validity of a mathematical system, that of decidability.

His technique was to construct a hypothetical machine, one very like a typewriter, but with a strip of paper replacing the ribbon, a mechanism for moving the strip in both directions one space at a time, a print head for printing marks on the strip and a scanner for reading marks off it. The machine could perform what sounded like quite sophisticated functions, such as recognizing particular patterns of marks, and looking up the pattern in a 'table of behaviour' to see what it should do next. In fact, the actual mechanisms needed to perform such functions could easily be implemented using standard mechanics – it would not be necessary even to use electronics. A fully operational Turing machine, were one to be built based on the design specification outlined in his original paper, would be of the same order of complexity as a typewriter or a chiming clock.

The 'table of behaviour', rather than the particular design of any components, was the key to the machine; indeed, it was the machine. For example, Turing showed that there was a table of behaviour that would turn the machine into an adding machine. Similarly, it could be turned into a machine that performed any of the basic operations of arithmetic. It followed, therefore, that there was a table, or set of tables, that could be drawn up to calculate any number that was calculable or, in Turing's chosen terminology, 'computable'. This table would perform the work without the need for any human intervention using a purely mechanical, 'algorithmic' procedure that would eventually yield a result. It was this hypothetical machine that enabled Turing to show that not all numbers could be computed, which in turn meant Hilbert's last criterion could not be met: there was no definite method that could solve every mathematical problem.

Strangely, even though Hilbert's project had been well and truly destroyed, and the work of logicians like Bertrand Russell had been abandoned, the underlying faith in the foundations of

Virtual Worlds

mathematics, the idea that it was somehow a pure expression of truth, persisted. It was more than just a 'game', because it submitted itself to the judgement of the most fundamental principles logic. Gödel and Turing based the validity of their arguments on logical principles: they had shown up contradictions, and logic does not allow such things.

Ludwig Wittgenstein, the Cambridge philosopher whose ideas were as wild as his eyes, was not impressed by this reliance on the tenets of a formal logic system. Three years after the publication of 'On computable numbers', Turing attended a series of lectures given by Wittgenstein on the philosophy of mathematics, in which the latter set about undermining Turing's faith in the necessity to discover 'hidden contradictions'. Hilbert had said: 'No one is going to turn us out of the paradise which Cantor has created', the paradise of mathematical purity. Wittgenstein responded: 'I wouldn't dream of trying to drive anyone out of this paradise. I would do something quite different; I would try to show you that it is not a paradise.'[9]

Turing, however, was convinced he was in Eden, which was why his mathematics had to remain uncontaminated from even the possibility of contradiction. Wittgenstein, on the other hand, wondered what damage contradiction could do. 'You cannot be confident about applying your calculus until you know there is no hidden contradiction there', said Turing. Wittgenstein replied that all contradiction would do would be to stop you applying the calculus, because it would simply fail to work. It would not produce a wrong result, it would not produce any result at all. The two kept up this ding-dong debate until Turing finally stopped attending. The fact that the issues they argued over were never resolved – or even, one could argue, clearly defined – was a symptom of a growing impatience with the need to consider the foundations or philosophy of mathematics. The debate seemed no longer to matter, and it dropped from the mathematical agenda. Only in the 1990s was it to return as a matter of significance, restored to its former position by the emergence of virtual reality.

In 1939, it was the world that mattered, not paradise. Turing's ideas were revealing themselves to have very practical significance. The idea of the Turing machine's tables of behaviour suggested that there was a way of formally describing any machine that could perform any computation: in other words, every computer is, in the end, a Turing 'table of behaviour'. This both put formal limits on what could be computed (there will never be the invention of some supercomputer that can compute the uncomputable) and, on the positive side, showed that it was possible to construct a machine to perform any computation.

The discovery that there could be a computer that could compute any computable number does not sound like the most shattering intellectual advance. But that is because we have got used to the idea of the computer. In 1936, it meant a person. Following Turing's insight, it meant a machine: he had proved, in other words, that it was possible to mechanize what had previously only been possible by means of mental effort. The machine had crossed a critical barrier. Before, machines had taken over the body, now they threatened to take over the mind.

What, though, did this machine produce? Nothing of material significance; just numbers. What use was a machine that just produced numbers? Obviously, it could be used to perform calculations. But the development of calculating machines had no need of Turing's theoretical designs. All the mechanisms required to perform any basic arithmetical function were well known by the late 1930s, so the building of a machine capable of computing any computable number was already a practical possibility. Turing had shown that some numbers were uncomputable, which was interesting, maybe even useful, but unlikely to change the course of technological history. Turing's insight, however, did much more than just establish the limits of mechanical calculation. It introduced the idea of the universal machine, a machine that can be lots of different machines; in fact, a machine that is capable of being any machine capable of performing a computation.

The idea of an abstract, immaterial machine is a difficult one, as paradoxical now as it must have seemed when it was first introduced. A mechanism needs moving parts, cogs and wheels, components, in order to work. So, it was perhaps unsurprising that when personal computers were first introduced to the shops, no one could quite work out what they would do. Writers, for example, started to express an interest in buying a 'word processor', by which they meant a machine for writing, editing and printing text that would replace their typewriter. They did not want a computer, even one that would run a word processing software package. So, to ease the confusion, a number of manufacturers started marketing computers as high-tech type-writers called word processors, even though they were, in fact, standard personal computers and could, like other personal computers, run other software turning them into different sorts of machines: machines for compiling and maintaining business accounts, for example. Such confusion was not the result of a lack of computer 'literacy'. No amount of IT training and hands-on experience can help. It was the result of what the philosopher Gilbert Ryle called a 'category mistake', a mixing up of things belonging to incompatible logical categories, like saying 'the universe is bigger than the number three'.

Category mistakes, though basic, are easy to make. A visitor to Oxford who, having been shown all the college buildings, asks where the university is, has made one. Ryle's aim in identifying this sort of mistake was, as we shall see, to exorcise what he famously described as the dogma of the 'ghost in the machine'. He wanted to show that the concept of 'mind' had the same confused relation to the body as the visitor's concept of 'university' had to the buildings he visited. Calling a piece of physical machinery a 'word processor' – or even a 'computer' – could be described as just such a category mistake. For practical purposes, it is a mistake that can be patched over, because any conceptual problems it might produce can be sidestepped by adopting the distinction between hardware and software. Hardware is the physical object, we are told, and software the list of commands that tell it what to do. Hardware determines certain characteristics, like speed, while software determines others, like function.

But the hardware/software distinction does not work all that well when applied with any theoretical rigour. A Turing machine, for example, can be simulated on another computer using software. But it can also be built (quite easily: all you need is some tape, a little circuitry and a couple of lamps) in hardware. And there is no difference between what the software and the hardware versions can do, they will be functionally identical. Neither would be very effective computers, but computers they would nevertheless both be.

There is a solution to this conceptual confusion, and it lies in the adoption of that word 'virtual'. A computer is a 'virtual' machine – a virtual Turing machine, to be precise. It is an abstract entity or process that has found physical expression, that has been 'realized'. It is a simulation, only not necessarily a simulation of anything actual.

Using a computer gives some experience of what 'virtual' really means. Personal computer users generally become comfortable with the idea of the system being at once a word processor, a calculator, a drawing pad, a reference library, a spelling checker. If they pulled their system apart, or the disks that contain the software, they would find no sign of any of these things, any more than the dismemberment of an IBM 370 would reveal all that extra memory provided by the virtual memory system. They are purely abstract entities, in being independent of any particular physical embodiment, but real nonetheless. 'Virtual', then, is a mode of simulated existence resulting from computation. Computers are virtual, not actual, entities.

Some readers may have noticed that in my attempts to reach down to the conceptual principles of computing, the one idea that remains unexamined is that of 'information'. Since computing is sometimes called 'information technology', this may seem to be a rather perverse omission. Much is made of the idea of information in discussions about the origins of computing. Specifically, Claude Shannon's and Warren Weaver's 1949 book *The Mathematical Theory of Communications* is cited. In developing his communications theory, Shannon used the concept of information to mean any communicated message, regardless of its meaning. This was an important insight, because it meant that it was possible to examine the problems of communicating a message, of distinguishing 'signal' (the original message) from 'noise' (interference), without resorting to semantics – what the sender of the message meant, and what the receiver of the message understood by it.

In *Mind Tools*, a book about the 'mathematics of information', Rudy Rucker wrote: 'the concept of information currently resists any really precise definition. Relative to information we are in a condition something like the condition of 17th century scientists regarding energy. We know there is an important concept here, a concept with many manifestations, but we do not yet know how to talk about it in exactly the right way'.[10] There is another possible explanation. Widened out by Shannon's defintion, the word is communicating less and less – we have lost the signal.

There is information 'space', information 'anxiety', information 'overload', even an information 'age'. In the first, we seem (this is just a rough guess) to be talking about something a bit like Hilbert space, in the second, a concern that we cannot deal with the amount of information from TV and newspapers, in the third the result of failing to heed our information anxiety, in the fourth, the observation that people talk about 'information' a lot these days. Money, phone calls, an architectural model, the smell of rose petals, pi, 'Stairway to heaven' by Led Zeppelin, DNA and the light of a distant star are all information. What does that tell us about them? Lumping them together in this way tends to serve more to hide than reveal important distinctions. For that reason, I suspect that information will not yield to definition because its meaning has become so indefinite. 'Information', wrote Theodore Roszak,[11] 'has taken on the quality of that impalpable, invisible, but plaudit-winning silk from which the emperor's ethereal gown was supposedly spun.' Look, and you find nothing there.

This is why I prefer words like 'virtual', 'abstract', 'mathematical' and 'computable'. We have to understand that a computer is a machine that performs a very precise function – mathematical computation. It can do no more. But that is a great deal, because there seems to be an underlying mathematical structure to everything that has been successfully analysed by science. 'Philosophy is written in that great book which ever lies before our eyes, I mean the universe,' wrote Galileo in 1623, 'but we cannot understand it if we do not first learn the language and grasp the symbols in which it is written. This book is written in the mathematical language, and the symbols are triangles, circles, and other geometrical figures, without whose help it is humanly impossible to comprehend a single word of it.'[12] This language is more than just information. It has the important and unique quality of being computable; it can be written by a machine. Maybe the universe is not a book so much as a computer, everything that exists within it the product of some algorithm. If so, this would mean that Turing's universal machine would truly be universal: given the right table of behaviour, and sufficient time, it could reproduce an entire virtual universe. Never mind flight simulators; how about world simulators?

Notes

1   Richard Sarson, 'A nominal way to achieve greatness', *PC Week*, 30 April 1991, p. 9.
2   Richard Thomas DeLamarter, *Big Blue: IBM's use and abuse of power*, London: Macmillan, 1987, p. 59.
3   Alan Turing, 'On computable numbers with an application to the entscheidungs problem', *Proceedings London Mathematical Society*, July 1937, 42, pp. 230–65.
4   Pierre Simon Laplace, *Essai sur les probabilities*, in Andrew Hodges, *Alan Turing: the enigma of intelligence*, London: Unwin Paperbacks, 1985, p. 64.
5   Stuart Hollingdale, *Makers of Mathematics*, London: Penguin, 1989, p. 312.
6   Hodges, 1985, p. 81.
7   In Hodges, 1985, p. 92.
8   Bertrand Russell, *Sceptical Essays*, 1928, in Tony Augarde (ed.) *The Oxford Dictionary of Modern Quotations*, Oxford: Oxford University Press, 1991.
9   Ray Monk, *Ludwig Wittgenstein*, London: Jonathan Cape, 1990, p. 416.
10 Rudy Rucker, *Mind Tools*, London: Penguin, 1988, p. 26.
11 Theodore Roszak, *The Cult of Information*, Cambridge: Lutterworth, 1986, p. ix.
12 In John D. Barrow, *The World Within the World*, Oxford: Oxford University Press, 1988, p. 238.

4
COMPUTABILITY

'Lacking a good French name for its devices,' wrote Ithiel de Sola Pool, 'IBM turned to Prof. J. Perret of the Sorbonne, who suggested the name *"ordinateur"*. That was a theological word which had fallen into desuetude for six centuries. "God was the great *ordinateur* of the world; that is to say the one who made it orderly and according to plan."' Did he? How could the professor be so sure?

The orderliness of the universe is a strange phenomenon, but one that has until now been a matter of purely philosophical debate. Why is it that the universe displays any order? Is it purely for humanity's convenience? Do we impose order upon the universe? Thanks to the *ordinateur*, such questions are no longer just academic. Is, as Professor Perret suggested in his choice of term, the universe a computation? If it is, then it must follow that any natural, physical process can be simulated using a computer, including life and human intelligence. Indeed, it implies that the entire universe could be simulated and our destiny calculated – an implication that some computer researchers, the 'computationalists', as I shall call them, believe to be true.

The history of science is not as simple as it once was. The old *Look and Learn* strange-but-true certainties have been eroded by the sophisticated arguments of theorists like Thomas Kuhn, the man who gave us 'paradigm shifts', Jean-François Lyotard, who saw science as a sort of convincing story, and Paul Feyerabend, a self-styled intellectual anarchist (or 'dadaist', as he later preferred to call himself). Nevertheless, we shall, for the moment, blinker ourselves from the implications of such arguments, and look at the origins of scientific progress as it might be related in *Natural*

Wonders Every Child Should Know, the book that Alan Turing had been given when he was 10 and which he told his mother opened his eyes to science.[1] In this account, science began with two geniuses, Galileo and Isaac Newton.

Galileo famously fell out with the Roman Catholic church for endorsing the Copernican view of the universe, which put the sun, rather than the earth, at the centre of the universe. It was this confrontation that asserted the independence of secular knowledge, truths built on the rock of experimentation and observation rather than the church. Galileo had, with his determined manner and his probing telescope, developed an approach to truth that has since become perhaps the most successful aspect of scientific practice: the insistence that only questions that can be answered should be asked. It was this approach that wrenched science from the clutches of philosophy. It was liberated from the imponderables. This is, of course, something of a caricature. Philosophers are not so stupid as to persist with unanswerable questions unless they know there is some value to be gained from the attempt. They are like international travellers, people who get more pleasure from travelling than arriving. Scientists, in contrast, see argument as simply a way of getting from A to QED.

The important lesson that every child should take from this is that science is not, as we suppose it to be, defined by its subject matter. It is a method, an approach, a tightly specified, rule-governed procedure – though a procedure that some argue does not have the monopoly on knowledge it claims for itself. Galileo is credited with helping to formalize this procedure, by firmly anchoring it to the principle of objective observation and loosening it from the principle of faith.

Newton's contribution was to show how astonishingly productive this method could be. In a way, he did for the dynamics of the physical universe what Euclid had done for its shape. He came up with a set of general principles or laws which could be used to predict the movement of bodies. These were abstract principles applied to ideal objects such as frictionless bodies. This meant that though they could never precisely predict the individual behaviour of real bodies, they could provide useful approximations for lots of different kinds of bodies, from falling apples to planets. Furthermore, Newton expressed these principles in a form that meant they could be applied mathematically. His second law of motion, for example, reads: 'A change in motion is proportional to the motive force impressed and takes place in the [same] direction of the straight line along which the force is impressed.' The concepts in this law – change, proportion, direction – are ones that can be formulated as mathematical relations and values, which means, for example, that it can be used to compute the movement of the planets.

Being so well acquainted with the idea of laws of nature, we can have difficulties appreciating what a stroke of luck or genius their formulation has been. There is no reason why we should assume that the complexity of experience has any order to it at all. Why should the movement of a planet have anything to do with the fall of an apple? Why should the force of gravity apply to both, in a way that can be worked out using the simplest of mathematical formulae?

A contemporary of Newton's, the soldier and man of letters Charles Boyle, the Earl of Orrery, had a mechanism named after him that reproduced the movement of the planets around the sun by means of clockwork. The orrery, which must have been as difficult to engineer as it is to pronounce, is now a highly collectable antique scientific instrument, though probably because it usually has such fine brass fretwork rather than because it embodied the change in world view that Newton's laws so precisely formulated. An orrery is an abstract simulation of the mechanical universe. The planets are brass baubles that revolve concentrically around the sun, each one in its proper place and each one at its proper speed. Orreries miss out many of the details of the solar system, including a few of its planets, but capture some of its more interesting characteristics, not least the strange movements of some of the planets as seen from earth, the phenomenon that led to the Copernican revolution.

Orreries are not particularly accurate, nor were they intended to be. They simply show that the planets move not according to a will of their own or the will of god, but according to certain theoretical principles. Furthermore, these principles are not just approximate guides, rules of thumb; they are the *governing* principles.

Demanding so much of the laws of nature – that they, in the words of Stephen Hawking speculating on the discovery of one single underlying principle, can lead us to 'know the mind of

God' – may seem foolhardy, even arrogant, but it is necessary to the whole conception of scientific reality. Nature without its laws would be truly incomprehensible. Every phenomenon could only be explained by describing it in complete detail. It would be like trying to teach chess by going through every single possible move, or trying to understand a language by learning every possible sentence it can express.

What, though, if there were no laws? What material difference would it make? Arguably, not a great deal, at least not until this century. Technology has not always been as closely coupled to science as is sometimes assumed. Much of it – Victorian engineering, for example – is based on an independent body of lore and experience. The industrial revolution might have even been possible without there first being a scientific one. Nevertheless, the fact that technology is not a constant process of invention seems pretty good evidence that there is some order in the universe.

Perhaps even more remarkable than this order is the fact that it seems to be mathematical. As noted earlier, Galileo regarded the book of the universe as a text written in the language of mathematics. This is what made Newton's laws more than mere observations and generalizations. They could compute the motion of bodies; they predicted the future.

* * *

As scientific theory has developed, so has the power of mathematics. It has proved itself capable of coping with anything scientists care to throw at it. There is no satisfactory reason for this; it simply works. And it is an article of scientific faith that it will continue to do so.

It is this faith that inspires the claim that the universe is a computation. To some extent, this is nothing more than an example of an opportunistic metaphor. The idea of the universe being a machine goes back at least to the invention of the orrery, and each age simply compares the universe with the latest type of machine. The computer is a current preoccupation, so everything that matters, the universe included, is seen as being some sort of computer. But the universe-as-computation is more than just a metaphor. If the laws of physics are mathematical, perhaps they are computable. Perhaps everything is in some mathematical relation to everything else. Since the universal Turing machine is capable of performing any arithmetical computation, then a Turing machine could, in principle, 'run' the universe. Put another way, perhaps the universe is really, not metaphorically, a Turing machine, a pattern of perpetual computation.

Thought about in detail, this idea is difficult to visualize. A Turing machine is easily imagined as a sort of typewriter or tape-recorder, with a strip of tape – albeit infinite – shifting this way and that as symbols are read and written. But the universe as a Turing machine? Where is the tape? What form does the table of behaviour take – is it contained in the laws of physics? One way of visualizing the universe as a computer is to use another dominant metaphor of our times – one that we seem to be drawn towards time and time again – that of the game. Using this metaphor, you imagine a much simplified universe, in fact one that is like a chess board, a simple, two-dimensional grid. Each square in this grid is a 'cell' that can have different types of pieces in it. The rules of the game govern what happens to the pieces. There are several weaknesses in this metaphorical universe game. To begin with, it is not a competitive game: there are no winners or losers. Second, it is not 'played' by anyone (except God, if you like). Third, there is no choice in the moves; the next move is determined by the former. However, the universe game does, as we shall see, have its metaphorical strengths, too.

Let us propose a simple set of rules by which a game can be played. These rules govern what happens to each square on the board. If any one square has two surrounding squares occupied by a piece (there is only one type of piece), it stays as it is, whether it is occupied by a piece or not. If, however, it has three occupied neighbours, a piece is placed in that square if there is not already one there. If it has more than three or less than two occupied neighbouring squares, then the piece is removed, unless it is already empty. These three simple rules cover all eventualities (they are, in the language of David Hilbert, 'complete'): the fate of every square from one move to the next is determined simply by counting up the number of occupied neighbouring squares. The rules, like the laws of physics, can also be expressed mathematically, or as a simple computer programme.

The game is 'played' by looking at each square and seeing what should happen to it, making the appropriate adjustments and then moving on to the next square, until the state of every square on the board has been calculated, whereupon you start all over again. It is not, as you might gather from this, much of an entertainment compared with other board games. However, because its laws are mathematical, there is no need to make the moves yourself. They can be performed by a computer.

When played at speed on a computer, strange phenomena start to appear on the game's 'board' (which, in the computed case, takes the form of dots, representing the pieces, appearing and disappearing on the screen, representing the board). With a sufficient number of judiciously scattered pieces on the board to start with, patterns quite unexpectedly start to appear: patterns like little groups of pieces – known by the game's enthusiasts as 'gliders' – that apparently 'fly' across the board. These gliders are not, of course, composed of the same pieces from one move to the next, just as an eddy that ripples around a rock in a stream is not composed of the same water from one moment to the next. But it is an identifiable ordered pattern that seems to form spontaneously out of the random spray of pieces.

In his book *The Recursive Universe*,[2] William Poundstone explained how this simple game could produce the most complex patterns, patterns that spit out gliders ('glider guns'), that form 'flotillas', that consume other patterns. There is even a pattern that performs the functions of a Turing machine, a universal computer, with gliders acting as individual 'bits' of data that combine and collide with different patterns to reproduce basic arithmetical operations.

Perhaps, though, the most exciting pattern that could emerge from such a game is one called a 'universal constructor', that is a pattern that can reproduce any other pattern, including its own pattern: it would, in other words, self-reproduce. The idea of the universal constructor was first developed by John von Neumann. The choice of term intentionally echoed Turing's idea of the universal computer. Von Neumann was responding to the growing postwar interest in automated manufacturing. Is there a machine, he wondered, capable of producing any other mechanism or manufactured product? Implicit in this question is a rather deeper and more puzzling one: is there a machine that can manufacture a copy of itself, including the mechanism needed to produce a copy of itself?

Von Neumann began by imagining the constructor as a sort robot swimming through a lake of spare parts, assembling the components it needs as it goes along. However, at the suggestion of Stan Ulam, a colleague at Los Alamos (where they had both worked on the development of the atomic bomb), von Neumann began to approach the problem along more formal, abstract lines using what were called 'cellular automata'. These created the same sort of game as the one described above though, in von Neumann's case, using many more types of pieces and much more complex rules.

Using this technique, von Neumann was able, like Turing, to examine the notion of self-reproducing machines by specifying a hypothetical, abstract one. The result was proof that, using the given rules and types of pieces, there was a machine, or rather a pattern, that could reproduce any other pattern, including a copy of itself. Astonishingly, however, it turned out that exactly the same could be achieved using the simple one-piece, three-rule cellular automata game described above. This demonstrated that all the universal constructor's complexity could result from the application of the simplest rules.

It was because of the success of these simple rules to produce such patterns that the man who devised them, the mathematician John Conway, called the game 'Life'. Even though he was said to have worked out the principles moving teaplates around a tiled kitchen floor, he quickly recognized that the simplest imaginable rules could produce unimaginable complexity. Having discovered the glider, Conway issued a challenge via the *Scientific American*[3] column that had first publicized his discovery: could anyone come up with a pattern than would actually produce gliders? A programmer called Bill Gosper at the Massachusetts Institute of Technology came up with the solution, the glider gun, and in the process formed an attachment to the exploration of the 'life universe', the incredibly rich realm of patterns produced by Conway's rules, that persisted for decades.

Gosper set the tone of the subsequent Life questing that overtook the imagination and computing resources of programming departments around the world. He first encountered the game while working on the now legendary ninth floor of MIT's Tech

Square building, a place where many of the programming greats of the computing revolution cut their first code.[4] But his interest did not lapse with the discovery of the glider gun or his departure from MIT. Working at night, using spare computing time donated by generous and indulgent employers, he, along with a community of Life-obsessed programming 'hackers' that flourished in the 1980s, embarked on an exploration of the exotic lifeforms produced by Conway's rules. The universe, he discovered, was one that naturally lent itself to the language of physics and even biology. It even provided a playful way of examining certain fundamentals of physics. The rate at which information passed through the Life universe in each successive move could, for example, be seen as the equivalent of the speed of light, an absolute, unerring constant that it is simply nonsense to imagine being flouted. Similarly, the game can be played on a 'board' with the top connected to the bottom and one side to the other, so a glider disappearing off the top of the computer screen reappears at the bottom, giving a perfect demonstration of the concepts of 'curved space' and a 'boundless but finite' universe that emerged from Einstein's relativity theory.

Life also had the attraction of generating a universe that appears to have a life of its own, at least in the sense that it can be left to develop of its own accord, producing results that could never have been anticipated from the starting position. The patterns moving around the screen are a bit like scurrying insects revealed beneath a lifted stone, or plankton in a rock pool. They are in a world of their own, and you can only wonder what that world is like and the purpose of their lives within it.

The idea of a flat universe like the one created by Life is not a new one. In 1884, Edwin A. Abbott, a teacher credited with the introduction of English literature as a school subject, published *Flatland*.[5] Described by its subtitle as 'a romance of many dimensions by A. Square', it postulated the existence of a two-dimensional universe in which all the creatures were flat shapes. A. Square was the narrator, a worthy, regular shape for a Victorian hero. He lived in a world populated by polygons of every conceivable shape, representing all the various classes of men. The women, however, were all needle-shaped, an entirely inappropriate choice of geometry, one might think, but intended to explain the reason why they had to enter all buildings by a separate entrance, because, viewed needle-point on, they were virtually invisible, raising the risk of the men sustaining fatal injuries by running into them. Since they had 'no pretensions to an angle, being inferior in this respect to the very lowest of the Isosceles', they were also, according to A. Square's testimony, 'wholly devoid of brain-power, and have neither reflection, judgement nor forethought, and hardly any memory. Hence, in their fits of fury, they remember no claims and recognize no distinctions. I have actually known a case where a Woman has exterminated her whole household, and half an hour afterwards, when her rage was over and the fragments swept away, has asked what has become of her husband and children'.[6]

Abbott clearly enjoyed the lack of inhibition that comes from writing about another universe – a liberation exploited by Abbott's contemporary Lewis Carroll in Wonderland. *Flatland* provided him with a chance to satirize the simplistic, arbitrary classifications of Victorian society. But the book also offers a demonstration of the difficulties of breaking out of the mental structures we use to make sense of the world. How would A. Square make sense of Spaceland, the three-dimensional world occupied by the strange Sphere creature that can be experienced in Flatland as a circle that could make itself large and smaller, and that could appear and disappear at will?

In a way, *Flatland* was a celebration of abstract thinking, its ability to take us beyond the limits of an empirical, believe-it-only-when-you-see-it viewpoint, a peculiar thing for a respected Victorian teacher to espouse, but timely, nonetheless, anticipating the abstractionism of the coming century. His ideas persisted, with the American logician Charles Hinton developing the exercise with *An Episode of Flatland*, published in 1907, and in 1965 the Dutch physicist Dionys Burger publishing *Sphereland*, an attempt to combine Abbott's and Hinton's world into one that would demonstrate the idea of the curvature of space.[7]

In 1977, Alexander Dewdney, a Canadian computer scientist, was inspired to embark on his own exploration of two-dimensional space. 'I was reading a popular work on cosmology', he wrote in 1984, 'and came across the familiar analogy which describes the expansion of our own three-dimensional universe in terms of a balloon whose two-dimensional skin continually expands. This led me to speculate about whether it would be possible, taking the balloon analogy literally, for a two-dimensional universe actually to exist. What sort of physics and chemistry would it have? What sort of life forms?'[8] Two years later, he published a monograph called *Two-Dimensional Science and Technology*, which Martin Gardner featured in his 'mathematical games' column in the July, 1980 edition of *Scientific American*, the very same column that had first introduced Conway's Game of Life to a wider public in 1970.

Martin Gardner has made a unique contribution to the opening of the abstract realms of Flatland and the Life universe to a growing number of programmers and hackers hungry to make sense of the power of the computer. His interest in the links between imaginary worlds and computer-generated ones was reflected in his *Annotated Alice*, which provided detailed notes on Lewis Carroll's Wonderland. Inspired by the thousands of letters he received in response to Gardner's well-read column, Dewdney decided to pursue the idea of a flat universe further, and eventually wrote *The Planiverse: computer contact with a 2-dimensional world*, a novel which followed the familiar conceit of presenting itself as a factual account. Dewdney (presenting himself as the 'compiler' rather than the 'author') wrote of running a program called 2DWORLD on his departmental computer from which spontaneously emerged a creature called YNDRD. Having discovered YNDRD, Dewdney and his students went on to uncover an entire world, complete with its own societies, cities and seasons.

Dewdney, who was himself later to write a *Scientific American* column under the title 'Mathematical recreations', was not proposing that the Game of Life could produce anything like the Planiverse or YNDRD. But he was exploiting, and perhaps strengthening, the link emerging in both the imagination and work of computer programmers and researchers between cellular automata 'games' and the physics of the real universe, a link that seemed to confirm that the universe might itself be a vast, three-dimensional cellular automaton – a very *big* automaton, with different rules and more pieces, but an automaton nonetheless.

The world around us is immeasurably more complicated than anything that could occur within the limited confines of computer 'game', but then the universe is potentially infinite in size and has existed for billions of years: it is a mother of a machine. But such an awesome difference in size does not particularly concern the computationalists: all they are interested in is establishing the principle. Have they, then, succeeded in this? One of the most compelling pieces of evidence that they might have is the connection between the patterns produced by programs like the game of Life and the origins of life itself.

* * *

Confronted by the difficulties of explaining why some things could display lifelike qualities and some could not, Aristotle came up with the less than conclusive argument that life was the result of 'spontaneous generation' – it just sprang up out of the earth. This was the ruling orthodoxy up until the study of anatomy began in Italy in the sixteenth century. In the early seventeenth century the English physician William Harvey followed up the studies of his teacher, the Italian anatomist Girolamo Fabrizio, into the generation of life by dissecting Charles I's hinds during the mating season to examine the contents of their uteruses.[9] Finding just a formless mass, Harvey concluded that the female must somehow be fertilized as a magnet magnetizes an iron bar that it touches. Refining his ideas, he came up with the concept of life arising out of organized matter, which he called the 'ovum' and by which he meant just about any substance – rotting meat and vegetation, dung – from which a living creature had been seen to emerge.

This conclusion may seem to be a classic example of prescientific thinking, the biological equivalent of flat earthism. But in many ways the more scientific accounts that emerged from subsequent work into the origins of life and the formation of living things seem to have increased rather than diminished the mystery. The work of Lazzaro Spallanzani and later Louis Pasteur, which demonstrated that life could only come from life, merely redefined the problem. The search for life's originating 'spark' still reached no satisfactory conclusion. In 1953, the American scientists Stanley Miller and Harold Urey mixed a 'chemical soup' with ingredients similar to those that would have been found in earth's primeval oceans which, energized by sparks representing lightning, reacted to produce the basic chemical compounds associated with living organisms. But this only proved the means by which the building blocks of life may have emerged, it did not show how they came to be assembled into something as complex as a living organism.

Perhaps Harvey was nearer the truth in seeing that it was the level of organization of matter that was key. How could the formless mass he discovered through the rather brutal demise of his king's bitches turn itself into a living creature? In 1943, the physicist Erwin Shrödinger gave a lecture at Trinity College, Dublin entitled 'What is life?'[10] In it, he described the chromosome as a 'codescript', thereby linking the idea of the formation of organisms to the idea of information rather than shape, a connection that was to lead to molecular biology.

Shrödinger also said that he thought living matter, while not independent of the laws of physics, must be the result of further, undiscovered laws.[11] His reason was that, unlike inanimate matter, living organisms maintain themselves in an ordered but unstable state, they do not, while they are alive, return to a state of equilibrium. Living things are in a constant state of organized flux – eating, excreting, inhaling, exhaling, growing, moving – a state that contravenes the principle that everything is drawn towards a state of shapeless disorder. While inanimate things are worn down, losing shape and form, living things somehow create it.

In a paper published in 1952, 'The chemical basis of morphogenesis',[12] Alan Turing addressed the issue of 'morphogenesis', the genesis of form, in terms of a mathematical simulation of what is known as a 'reaction-diffusion' model. Rather than look for new laws of physics to explain life, he used the model to reproduce the interaction of chemicals in a formless soup. What he discovered, using a program he had developed for one of the world's first electronic computers at Manchester University, was that patterns spontaneously emerged from his simulated soup, patterns that were in no way evident before the soup was allowed to 'cook'.

The experiment was a classic example of the power of simulation and modelling. Previously, reaction-diffusion models would not have been much use, because they are based on 'non-linear' rather than 'linear' mathematics. The linear relationship between two quantities is a very easy one to work out, it means they are related in direct proportion to each other. A linear relationship is expressed in phrases such as 'the more you put in, the more you get out'. It is plotted by a straight line on a graph. A non-linear relationship is not so straightforward – literally. It is not a straight line on a graph, but a curved one. Exponential growth, the population explosion, inflation, feedback, nuclear melt-down, these are all the products of non-linear relationships. The problem with the non-linear equations is that they require an enormous amount of time to calculate. Working out a non-linear equation is a bit like trying to work out a seating plan for a dinner party at which most of the guests dislike all but one or two of the other guests. Adding each new guest to the plan usually means recalculating the positions of every other guest.

Without a computer, Turing would never have been able to capture the dynamics of a reaction-diffusion model; to revive an earlier analogy, trying to work out what was happening on the basis of even hundreds of hand calculations would have been like trying to understand the narrative of a film by looking at a few of its frames. Only with the computer could the organization of the 'soup' into interesting patterns be observed. The computer, like a new scientific instrument, had revealed something that could never have been seen before: the spontaneous generation of form. It had shown that morphogenesis was not necessarily the result of some external, mysterious 'force' that acted on the physical material of life, but a quality of form itself.

Turing's work was later to translate directly into the language of cellular automata. One of the most impressive classes of cellular automata is one that has been dubbed 'Hodgepodge', after a set of rules developed by Martin Gerhardt and Heike Schuster at the University of Bielefeld in Germany. In Conway's game, cells (the squares on the board, to revert to the board game analogy) had one of two states, dead or alive (the presence or absence of a piece). In Gerhardt and Schuster's version, they could have a number of states, one being, say, 'healthy', the others varying degrees of illness. Furthermore, since the state of a cell's health is set by the state of its neighbouring cells, illnesses 'spread' (this epidemiological language should not be taken too seriously; it is simply a way of explaining the rules).

The result of running Hodgepodge is a classic reaction-diffusion model. Furthermore, it is one that produces results that are strikingly similar to a class of real chemical reactions rather forbiddingly known as Belousov-Zhabotinsky reactions. These are typical examples of 'self-organizing' phenomena, and the ability to reproduce exactly the same sort of organization using cellular automata seems to confirm that there must be some connection, some underlying organizational mechanism that operates in the same way in both the Hodgepodge games and the Belousov-Zhabotinsky reactions. Also, other research has shown how, to borrow Kipling's phrase, the leopard got its spots, and whose immortal hand or eye, to borrow William Blake's, framed the tiger's fearful symmetry – it was the pattern-forming processes of a reaction-diffusion model.

Such results made life a perfect test case for the computationalists, because it turned the argument about the 'mystery' into one about order and complexity. It posed the problem in terms of a yet more fundamental one: what are the origins of the structure and organization in the universe? Why is it not simply random chaos? This was a question to which computer models such as cellular automata seemed to suggest some sort of answer. At least, so it seemed to Stephen Wolfram.

Wolfram was a recipient of a MacArthur Foundation 'genius' grant, a no-strings five-year bequest running into hundreds of thousands of dollars that is bestowed upon anyone the foundation regards as showing sufficient scientific promise to earn it. 'If you think about the thermodynamics of the early universe', said Wolfram, 'you get into a strange problem. The universe is supposed to have started out as this uniform ball of hot gas, but in the end what we see is a lot of galaxies that are very patchy and irregular. The question is how do you get one from the other? Standard statistical mechanics says that you can't.'[13] Cellular automata seem to say that you can, which is why Wolfram has devoted much of his generously funded time to exploring the way they work.

Wolfram's work, in combination with the more adventurous speculative explorations of people like Gosper and another veteran of MIT's Tech Square, Ed Fredkin, began to see cellular automata as providing evidence that certain combinations of simple elements 'self-organize' themselves; the structure simply pops out of the chaos. It does not sound like much of an explanation, more of a question-begging observation. But it is hard to deny that it happens, and hard not to be impressed that it happens.

Cellular automata and reaction-diffusion models are part of a much larger and well-publicized intellectual and scientific movement, one identified, thanks largely to James Gleick's book of the same name,[14] by the generic term 'chaos'. The Russian-Belgian Nobel laureate Ilya Prigogine co-wrote a book in 1984 which developed the idea of self-organization to show how the second law of thermodynamics, the law of 'entropy' which says that everything tends towards a state of disorder, can, without being violated, be shown actually to produce order.[15]

But, as other computer science researchers and mathematicians, mostly in America, have noticed, chaos does not just produce order, it *has* order: there is a deep structure – to borrow a phrase of linguistics – in the apparently random, chaotic behaviour that characterizes all natural and some social phenomena. This structure takes the form of what has been enticingly termed a 'strange attractor'. An attractor is a state towards which a system is drawn. A familiar example is the point of rest of a pendulum. A 'strange' attractor is an invisible state that comes from analysing very complex data, such as weather patterns, as was done in a famous experiment conducted by Edward Lorenz. Strange attractors are shapes that appear in phase space, in the realm of mathematics; you can never spot them in the physical world as you can the resting position of a pendulum. The exciting feature of them is that they suggest that perhaps all chaotic, apparently meaningless phenomena are drawn towards a strange attractor. For example, the unpredictable bull charges and bear hugs of stock markets may not, as is assumed, be the uncomputable result of lots of individual decisions taken by independent stock brokers, but the signature of a strange attractor. Of course, if anyone could, by analysing the stock market figures for a period of time, discover this strange attractor, they would be able to predict how it will change in the future.

On the debit side, though chaotic systems may be stable at the abstract level of the strange attractor, they are highly unstable at the level we experience them directly. Non-chaotic, linear systems exhibit gradual change; chaotic, non-linear ones fly off the handle at the slightest provocation. This is known as the 'butterfly effect'. The butterfly effect is the dramatic change in the behaviour of a chaotic system that can result from the tiniest changes in 'initial conditions' – how the breeze of a butterfly fluttering across an

English meadow might be enough to trip the earth's atmosphere from one state into another, from a state where there are snow-storms in Moscow to one where there are hurricanes in the Gulf of Mexico. The butterfly effect is quite well expressed in the phrase 'the straw that broke the camel's back', and captures the un-predictability of non-linear systems – the fact that the effect of one factor over another is not simply proportional. This would imply that any gains that come from knowing the stock market's strange attractor have to be balanced against losses that result from an arbitrarily tiny decision taken over some inconsequential stock producing an unanticipated crash.

Benoit B. Mandelbrot, nephew of the Bourbakist Szolem Mandelbrojt, sees himself, and is seen by many, as the champion of the more intuitive, less reductive set of scientific attitudes that are associated with chaos. Unlike his colleagues in university departments, he was able to cultivate an open mind to the 'mess', as he calls it, of nature. Whereas physicists since Newton endeavoured in their experiments to exclude this mess – dreaming of friction-free worlds and perfectly spherical particles – Mandel-brot wallowed in it. 'One should not say that the mountains are imperfect; they have their perfection',[16] he said.

He noticed in a variety of turbulent and irregular phenomena a 'self similarity' which would yield to mathematical analysis. He found that comparing the shape of mountains, clouds and plants across different scales (a branch and all its twigs, for example, looks much like the tree and all its branches) could form the basis of a new 'fractal' geometry, one that replaced lines and circles with bumps and lumps. As he once explained to me between mouthfuls of an art gallery cafe's piquantly-dressed leaves of highly fractal endive: 'I did not discover the fact that clouds are like billows upon billows upon billows. Every child knows that. Every landscape painter knows that. What I did was identify tools that turn this intuitive perception of shape into something that science can grab.' This was an achievement that demanded a radical adjustment in conventional mathematical attitudes. As James Gleick put it in his book *Chaos*: 'Unlike most mathematicians, [Mandelbrot] confronted problems by depending on his intuition about patterns and shapes. He mistrusted analysis, but he trusted his mental pictures.'[17]

The most extraordinary picture to result from this process was the Mandelbrot set. Generated by the simplest of formulae (take a number, square it, add the starting number, square the result, add the starting number, square the result and so on), the set graphically plotted an image of astonishing complexity that displayed all the features of self-similarity that Mandelbrot had found elsewhere. It seemed to hold some sort of universal meaning, plotting the boundary conditions that govern the behaviour of any number of potentially 'turbulent' or chaotic phenomena: stock exchange indices, galactic clusters, electrical interference. People couldn't – and still cannot – prevent themselves from talking about it as they would a painting, as an image revealing some great but mysterious truth.

Since Mandelbrot first produced a detailed picture of what he then called an 'island molecule' in 1980, there have, predictably, been several claims to its independent discovery.[18] Even ignoring the unseemly rows that accompany most high-profile discoveries, one of the interesting features of Mandelbrot's growing mythological status is that it echoed almost precisely the language used in connection with the French mathematician René Thom, who is credited with discovering 'catastrophe theory' in the 1960s.

Catastrophe theory was itself described in terms that anticipated Gleick and his like. '[Catastrophe] theory is controversial because it proposes that the mathematics underlying three hundred years of science, though powerful and successful, have encouraged a one-sided view of change. These mathematical principles are ideally suited to analyse . . . *smooth*, continuous, quantitative change: the smoothly curving paths of planets around the sun. . . . But there is another kind of change, too, change that is less suited to mathematical analysis: the *abrupt* bursting of a bubble, the discontinuous transition from ice at its melting point to water at its freezing point.'[19] Those words were written by Alexander Woodcock and Monte Davis in 1978, a year before Mandelbrot had the first inkling of his set, and only three years after the first published use of the word 'chaos' in anything like its scientific meaning.[20] Yet, substitute the word 'change' with, say, 'behaviour', and you have what could pass as a classic formulation of chaos theory. Quoting Thom, Woodstock and Davis characterize catastrophe theory as being a science of the 'ceaseless creation, evolution and the destruction of forms', an elegant expression of some of the preoccupations of chaos theory if ever there was one.

This does not mean that chaos theory somehow plagiarized catastrophe theory, or is engaged in some conspiracy to deprive it of its proper position in intellectual history – the ideas surrounding both emerged around the same time. Catastrophe theory is, furthermore, wholly concerned with a different type of mathematics, topology, the study of what has been called 'rubber-sheet geometry'. Topology is a tricky concept, its concern being every aspect of an object's form not dealt with by geometry. This is usually defined for non-mathematicians as dealing with those features of an object that are not changed by bending, stretching, compressing or twisting. So, for example, a pair of gloves may be the same shape and size, but are topologically different, because no amount of bending, stretching, compressing or twisting could turn a left-handed glove into a right-handed one. A square, on the other hand, is topologically identical to a circle, because you can bend one into the shape of the other.

To make the situation even harder to grasp, in catastrophe theory, Thom was not concerned with the topology of familiar physical objects, or even familiar abstract ones such as maps, but with the sorts of forms one finds in the mathematical realm of graphs, in that sort of abstract space opened up by Hilbert, and where the strange attractor was discovered. Thom's breakthrough was to analyse 'discontinuous change', the sudden movement of a system from one state to another (the freezing of water, the snapping of a stick, the swarming of locusts) by examining the topology of a graph that plots their movement from one state to another.

Catastrophe theory, therefore, was no more concerned with our everyday idea of catastrophe than chaos theory is concerned with our everyday idea of chaos (if anything, one could argue that both were concerned with the opposite of what their names suggested, with the discovery of pattern and order). But, as with chaos, the applications of the theory had an enticingly real-world ring to them. Thom's goal, like Turing's, was to develop a theory of morphogenesis, a way of accounting for the generation of form – his book on the theory was entitled *Structural Stability and Morphogenesis*.[21] But it was a theory that promised to help explain more pressing phenomena in an increasingly catastrophic century. Maps were drawn of population explosions and economic cycles to show how they shared common topographies – just as chaos theorists have studied the very same phenomena to see if they are governed by the same strange attractor.

Gleick argues that topology prevailed to the neglect of the mathematics of chaos.[22] Certainly, topology is regarded as one of the great preoccupations of twentieth-century mathematics, but the implication that catastrophe theory was part of an intellectual movement that stifled the early development of chaos theory seems uncharitable. Even less charitable is the fact that catastrophe theory was excluded from nearly every history of chaos, despite the fact that it has so much in common – Thom's theory even has a 'butterfly catastrophe'.

What, then, happened to catastrophe theory? Once an essential topic of the 1970s intellectual dinner-round, the 1980s saw it simply disappear from the agenda. The reason may partly be the character of Thom himself, a bit of a Mandelbrot, celebrated for his intuition and imagination, and in equal measure suspected for a lack of precision or thoroughness. He was impatient with those who insisted on 'formal rigour' – that could always come later, he claimed. He was one of those thinkers who is cast in the role of both a hero and rascal, inspirational to some, undisciplined to others, the object of intense media interest and intense professional jealousy.

But the problem with catastrophe theory was more than a matter of style. At its height, it claimed to provide a means of understanding any type of catastrophe, from prison riots through the outbreak of the First World War to schizophrenia. There seemed to be no limits to its application; it almost worked too well. A theory that has something to say about everything may end up saying nothing about anything. In 1977, the American journal *Science* ran an account of an attack on the theory levelled by the American mathematician Hector Sussmann under the headline: 'Catastrophe theory: the emperor has no clothes'.[23] Later that year, Sussmann published a paper with R. S. Zahler in *Nature* which concluded that 'the claims made for the theory are greatly exaggerated and its accomplishments, at least in biological and social sciences, are insignificant . . . Catastrophe theory is one of many attempts that have been made to deduce the world by thought alone . . . an appealing dream for mathematicians, but a dream that cannot come true.'[24]

One could only too easily imagine exactly the same sorts of

concerns producing a backlash against chaos. That, too, seems universal in its application – you start to see fractals everywhere, imagine that everything must be an expression of strange attractors hidden behind yet higher orders of abstraction. It, too, smacks of an attempt to 'deduce the world by thought alone', an example of French-inspired rationalist 'armchair science' that Anglo-Saxon culture so despises. Most significantly of all, chaos theory, like catastrophe theory, relies on the assumption that the real world and the mathematical world are in some way connected. It is only too tempting to accept this question as settled. But most of the 'discoveries' of the chaos that attracted scientists' attention were purely about the properties of a certain class of mathematical equations. Fractal geometry was described rather grandly by Mandelbrot as the geometry of nature, but there is no proof of this. Nature remains insolently silent on the matter. She will not say where she got her designs from.

With the passing of catastrophe theory and the no-doubt imminent lapse of interest in chaos, a new theory has emerged, one that could be called self-organizing criticality, but which one waggish science journalist anticipated being called crisis theory.[25] Crisis theory is, as its more prosaic scientific name suggested, a combination of the concept of criticality and self-organization. Prosaic though the word may be, criticality is a powerful notion. It is better known when used in connection with nuclear energy generation, and has filtered to some extent into everyday language in the phrase 'critical mass'. 'Criticality' lies at the edge of catastrophe. Nuclear power stations are held at precisely this point, the point where the release of energy is not too much to precipitate a run-away chain reaction – turning the core into a potential, slow-burn atomic bomb – nor too little to produce sufficient heat to boil the water that will drive the turbines.

The developers of the concept of self-organized criticality, led by Per Bak at the Brookhaven National Laboratory in the US, believe that the most interesting feature of large, unpredictable events – the same sorts of events that concerned the catastrophe and chaos theorists; stock market crashes, earthquakes, that kind of thing – are in fact a set of independent smaller events. An earthquake is the result of many slippages between moving plates in the earth's crust, a stock market crash is the result of lots of local decisions taken by individual investors. The fact of such systems or events having this 'multi-component' nature draws them, Bak and his colleagues argue, into a state of criticality, but they are drawn there not by any external force, but simply as a result of their multicomponent nature – they organize themselves, hence self-organized criticality.

Like Thom, Bak is concerned with explaining why systems that are on the point of crisis somehow manage to persist. The universe is supposed to be headed in a direction of equilibrium, smoothing itself down from hot peaks and frozen troughs to an average level of uniform warmth. Yet there are, at least around us on earth, all these structures, ourselves included, teetering on the edge of crisis (think how little it takes to kill a living creature) that are nevertheless stable, even self-sustaining. A more mundane example explored by Bak is a pile of sand which, despite being made up of nothing more than independently acting grains, nevertheless always somehow manages to organize itself into a neat heap, obligingly setting off little sand-slides when more sand is added to ensure that the shape is retained.

Bak attempted to explore the idea of a self-organizing sand pile's critical state by building a computer simulation of one with uniform square grains. This fairly accurately managed to model the way a sand pile behaved as grains are added, although researchers at the Cavendish Laboratory in Cambridge with real sand piles (such a lovely image; a group of Britain's top research scientists playing in a sand pit) discovered that 'natural angle for the slope of a [real] pile depends on how the pile was made', a phenomenon not found in the computer simulated pile.[26]

These three theories – catastrophe, chaos and crisis – have all succeeded in using mathematics to explore phenomena that are very *natural*. Mathematics is conventionally thought of as being a realm apart as much by the heathen outside as the fundamentalists within. Its teachings are not of this world; we all know it. No longer does such a view seem sustainable.

* * *

At the SIGGRAPH conference in Boston in 1989, a pile of leaflets was to be found nestling among the foothills of what is always a mountain of press information. It attracted my attention because it baldly proclaimed that there would be a demonstration of what was called a 'video modem'. I did not take it very seriously, since everyone, at least everyone attending such a conference, knows that such a device is a technological impossibility. They know that a modem is a device for translating computer information into a form that can be sent down telephone lines, and they know that this means, telephone lines being what they are, that paltry amounts of data can be sent, a few thousand bits a second, no more. They also know that a single frame of video, one of the 25 or 30 that make up each second of a moving TV image, is several million bits' worth of information. So no device could possibly send moving video images down the line.

Nevertheless, that is what the manufacturers claimed to have achieved – well, almost, allowing for the usual low tolerances of accuracy expected in any computing industry announcement. The trick was performed by Michael Barnsley using a process he called the iterated function system, IFS. IFS was based on the principle of creating pictures according to the self-similarity of their elements, the similarity of the twig to the branch, of the branch to the trunk. By analysing an image using this method, he discovered that most if not all the picture information could be reduced by a factor of a hundred thousand, sometimes by as much as a million. This meant images that would normally take up billions of bytes to store could be reduced to just a few hundred.

This was a technical breakthrough. Computer researchers are obsessed with finding yet more ingenious ways of 'compressing' information, cutting out any redundancy or repetition. But it went much deeper than that. The IFS system only worked because the world is fractal. During a demonstration of the system, Barnsley showed me his system reconstructing pictures of his 'girls', pictures of what looked like Taiwanese air hostesses, and of lush landscapes, from just a few hundred bytes of information stored on a floppy disk. These were images that had been compressed without any knowledge of their content: they had been analysed by the computer as simply a meaningless map of bits. And what they demonstrated is that all the richness, variety, all Mandelbrot's 'mess', had been shown to be not complex at all, but just a few simple shapes repeated over and over again on different scales. What Shakespeare called nature's 'infinite book' can be reduced to just a few characters. This seems to tell us two things. That nature is far simpler than it looks. And it tells us that a purely computational procedure – pure maths, in other words – can decode it, can automatically, mechanically discover its structure. You cannot watch Barnsley's demonstration and not be convinced that, unproven though it may be, maths must have some very intimate connection with the world and that the computer is capable of capturing that maths. More generally, fractals, cellular automata, chaos, catastrophe and a host of other mathematical developments seem to show that the world, nature, even life itself may, for all their messiness and complexity, be computable.

There is, however, one remaining test that we have yet to examine, taken by many to be the definitive test, the one that the computationalists must expect their beloved machine to pass if they are to argue with any credibility that the universe is a computer. That test is the challenge of artificial intelligence.

Notes

1 Andrew Hodges, *Alan Turing: the enigma of intelligence*, London: Unwin Paperbacks, 1985, p. 11.
2 William Poundstone, *The Recursive Universe*, Oxford: Oxford University Press, 1987.
3 Martin Gardner, 'Mathematical games', *Scientific American*, October 1970.
4 Stephen Levy, *Hackers*, New York: Dell, 1984, p. 144.
5 Edwin A. Abbott, *Flatland*, London: Penguin, 1986.
6 Abbott, 1986, p. 26.
7 A. K. Dewdney, *The Planiverse*, London: Pan, 1984, p. 268.
8 Dewdney, 1984, p. 268.
9 François Jacob, *The Logic of Life*, London: Penguin, 1989, p. 53.
10 Erwin Schrödinger, *What is Life?*, Cambridge: Cambridge University Press, 1944.
11 Walter Moore, *Schrödinger*, Cambridge: Cambridge University Press, 1989, p. 399.
12 Alan Turing, 'The chemical basis of morphogenesis', *Phil. Trans. Royal Society*, Series B: Biological Sciences (237), pp. 37–72, 1952.
13 In Ed Regis, *Who Got Einstein's Office?*, London: Penguin, 1987, p. 232.
14 James Gleick, *Chaos*, London: William Heinemann, 1988.
15 Ilya Prigogine and Isabelle Stenger, *Order out of Chaos: man's new dialogue with nature*, New York: Bantam, 1984.

16 Benjamin Woolley, 'Teletexts', *Listener*, 25 May 1989.

17 Gleick, 1988, p. 84.

18 John Horgan, 'Mandelbrot Set-To', *Scientific American*, April 1990, p. 14.

19 Alexander Woodcock and Monte Davis, *Catastrophe Theory*, London: Penguin, 1991, p. 9.

20 See Heinz R. Pagels, *The Dreams of Reason*, New York: Bantam, 1989, p. 80.

21 René Thom, *Structural Stability and Morphogenesis: an outline of a general theory of models*, Reading: Benjamin, 1975.

22 Gleick, 1988, p. 46.

23 G. B. Kolata, 'Catastrophe Theory: the emperor has no clothes', *Science*, 15 April 1977.

24 R. S. Zahler, 'Claims and accomplishments of applied catastrophe theory', *Nature*, 27 October 1977.

25 *The Economist*, 12 May 1990, p. 85.

26 Anita Mehta and Gary Barker, 'The self-organising sand pile', *New Scientist*, 1773, 15 June 1991, p. 40.

5
ARTIFICIAL INTELLIGENCE

A computer researcher in conversation with a journalist reveals that his area of research is 'artificial intelligence'. The journalist expresses some scepticism about this; intelligence seems to be something beyond the scientist's pale, part of the mysterious machinations of the mind. 'What', replies the researcher, used to such arguments, 'is so special about the mind?' *Mind*, the 'secret working mind', as Yeats called it; surely it does not need to prove itself special? It is the origin of our identity, of our consciousness, of our lives.

Science, as we have already discovered, is outrageously demanding. It demands that it is not simply a way of explaining certain bits of the world, or even the local quarter of the universe within telescopic range. It demands that it explains absolutely everything, or, more accurately, that its theories apply to everything relevant to them. A theory regarding the behaviour of molecules therefore relates to every single molecule in the universe, without one single exception. Should an exception be discovered, it would not be the exception that proves the rule, it would prove the rule false. The entire universe in its every detail and across all time obeys the laws of nature, according to science; the problem is discovering what those laws are – and even the scientific establishment is prepared to admit the laws it currently assumes to be correct are in all probability incorrect. However, the scientific voyage of discovery does have a well-developed sense of direction: towards the eventual discovery of the true laws of nature.

There has, however, been one region of doubt, one part of the universe that, perhaps, does not obey any such laws, and that is the human mind and all its subsidiary mysteries: consciousness, the self, identity. This is the assumption of what is known as 'dualism'. *Dualism* divides the universe into two types of substance, material and mental. The laws of nature only apply to material substance. Mental substance acts according to a will of its own, human free will.

The name most closely associated with dualism is Descartes, the man from whom, despite living in the first part of the seventeenth century, 'modern' philosophy is said to derive. 'There is a great difference between mind and body, in that body, by its nature, is always divisible and that mind is entirely indivisible', he wrote.[1] By making this distinction, not only was the will left to be free, but the world was left to be real, to follow laws rather than the whim of the imagination. In other words, Cartesian dualism is not about reducing the power of science and mathematics, but asserting that it can apply to the entire material universe. Our minds, meanwhile, can be left to make their free choices, choices for which we are responsible, choices which we have not been forced to make; we cannot, then, have those choices being the result of the mechanical laws of nature.

Cartesian dualism came under the most convincing – and entertaining – assault in 1949 by the Oxford philosopher Gilbert Ryle. Ryle was a friend of Ludwig Wittgenstein. The two seem to have been both drawn towards and repelled by each other. Ryle began by admiring Wittgenstein, Wittgenstein's attention was attracted by Ryle's apparently serious and interested expression during the reading of a paper. Later Ryle decided that Wittgenstein had a bad influence on his students, and Wittgenstein decided that Ryle was not serious after all. The two, nevertheless, remained cordial, being as likely to argue about films (Wittgenstein's contention being that a good British film was an impossibility) as philosophy.[2] Wittgenstein, more to our current point, can be seen as having had a powerful influence over the philosophical climate of the times, he was the prevailing wind, one that was blowing away many of the old ideas about the old issues.

Ryle seemed to be responding to that climate with the publication in 1932 of an article entitled 'Systematically misleading expressions'. In it, he introduced the idea of a 'category mistake', a concept that was to play an important part in his great work, *The Concept of Mind*, published in 1949.[3]

A category mistake is usually just a simple error, but one that can be hard to detect and correct. It is like woodworm that burrows through the argument's logical structure, invisibly weakening it, only identified too late, when the tiny holes reveal the infestation's departure. To explain what he meant by the term, Ryle gave a selection of rather quaint examples, reflecting, perhaps, the man's somewhat tight orbit of worldly experience (apart from war service and the occasional foreign trip to give lectures, he spent his entire working life in Oxford). A 'foreigner' is taken round the colleges, libraries, playing fields, museums, scientific departments and administrative offices of Oxford University. The tour completed, the visitor asks 'But where is the university?' His mistake lay in what Ryle kindly called the 'innocent assumption' that the university belonged to the same category of institutions that he had been shown during the tour. The tolerant guide, perhaps after referring the bemused visitor to Ryle's work, decides to take his guest to see a game of cricket. The rules of the game and the role of each player are patiently explained to him in the simplest words, but still he is puzzled. 'There is no one left on the field to contribute the famous element of team spirit', he says. Wrong again. Different category (different era, too).

As we saw earlier with the case of personal computing, our lives are filled with people making category mistakes. Take 'the voter'. Politicians wilfully ascribe all sorts of attributes to this 'voter' – intelligence, political discernment, impatience with the party in power/opposition – which it is quite impossible for an abstract concept rather than a physical human being to have. Given that no politician could have possibly met 'the voter', he shows a rude presumption of the intimacy of his acquaintanceship. Another is the song in the Disney cartoon *Dumbo*, featuring a barbershop quartet of crows singing about how they had seen a house fly, an elastic band, a baseball bat, but they had seen just about everything when they had seen an elephant fly.

In *The Concept of Mind*, Ryle argued that just such an error is made in the way we talk about the 'mind'. 'There is a doctrine', he wrote, 'about the nature and place of minds which is so prevalent among theorists and even among laymen that it deserves to be described as the official theory.' This theory was Descartes' 'myth' of dualism. 'With the doubtful exceptions of idiots and infants in arms every human being has both a body and a mind. . . . Human bodies are in space and are subject to the mechanical laws which govern all bodies in space. . . . But minds are not in space, nor are their operations subject to mechanical laws.' Ryle's attack on this 'official theory' was that, in a way, we believe our own language. We talk about the mind in such a way as though it is in the same category of things as the body, and from this deduce that there are two worlds, the mental and the physical. In his splendid, much-used headline phrase, we have created a 'ghost in the machine'. In fact, argued Ryle, when we talk about the mind, we are not talking about a separate 'world', we are simply using that sort of language to describe the property of a particular way we or others behave.

Ryle's attempt to exorcise the ghost was typical of his time, a time when the cosy Cartesian consensus was under the sustained bombardment. A decade earlier, the American psychologist B. F. Skinner had published *The Behaviour of Organisms*, which set out a powerful case for behaviourism, the idea that mental states are just behaviour. One of Skinner's most important experimental contributions was a device called the 'Skinner box', a device in which the subject (usually a rat) was placed to see how it would learn responses to, say, pressing a lever in order to receive food. Behaviourism was an attempt to turn psychology into 'real' science, a science that was only about observable and, preferably, measurable phenomena.

The important point about Ryles's argument and work by psychologists such as Skinner was that they dragged the mind out from behind the veil of dualism into the domain of scientific reality.

* * *

As early as 1946 the press was describing the ENIAC computer at the University of Pennsylvania as an 'electronic brain'.[4] At that time, the word 'computer' was still used to mean a worker engaged in calculation, and the very fact of its application to a machine clearly suggested a connection between the mental work of the human and the electronic processes of the machine. So, too, did the use of the word 'memory' for computer storage.

Alan Turing himself had assumed a link between computers and brains since he started to consider the concept of computation.

At the end of the Second World War, he told a friend that he intended to 'build a brain'.[5] By this, he did not mean a mechanical substitute for the spongy organ in the human head; he meant a machine that would capture what he assumed to be the formal characteristics of the brain, those characteristics to which Ryle would have said the concept of the mind actually referred. 'To understand the Turing model of "the brain"', wrote Turing's biographer Andrew Hodges, 'it was crucial to see that it regarded physics and chemistry . . . as essentially irrelevant. In his view, the physics and chemistry were only relevant in as much as they sustained the medium for the embodiment of discrete "states". . . . The claim was that whatever a brain did, it did by virtue of its structure as a logical system.'[6]

In 1933, the Hungarian-born chemist Michael Polanyi, having resigned from the Kaiser Wilhelm Institute in Berlin in protest at the dismissal of Jews, came to Manchester University as professor of physical chemistry. Given his recent experience, his growing preoccupation at Manchester with the relationship between individual freedom and science was understandable. He distrusted scientific orthodoxy and materialism, and was only too keen to accept Gödel's Theorem demonstrating the incompleteness of mathematics as evidence that there was room in the universe for the non-mechanistic.

In 1948, the year Polanyi was made professor of the then new field of social studies at Manchester, Turing was appointed deputy director of the university's newly formed computing department. While there, he debated the issue of mechanizing the mind with Polanyi. The following year, both of them joined a mixed bag of academics at a meeting held at the university on the theme of 'The mind and the computing machine'. Following this meeting, Turing wrote up his views in his 'Computing machinery and intelligence' paper,[7] and set computing on its quest to develop artificial intelligence.

Turing's paper was not an outline of any particular technology. Being written for a philosophical journal, its concern was more with principles than possible practice. Turing wanted to establish that there is no essential link between intelligence and humans. He did this by means of a game, one that he first imagined as a gender game – a significant choice, given the 'game' Turing, who was homosexual, was forced to play with prevailing social attitudes towards his sexuality. The game was simple. The player would interrogate two people in separate rooms. One person was a woman, the other a man. Both would, through written answers to questions asked by the player, try to prove that they were women.

The reason for the choice of game was that the difference between the hidden man and the woman is not dependent on what they actually write. If the man convinces the questioner that he is a woman by writing strikingly feminine responses to the questions, it would neither prove that the man was, in fact, a closet woman, nor the woman a closet man. However, if one of the participants was replaced by a computer, and the questioner continues to play and still thinks that the replies are coming from a human of either gender, that *would* prove something of the machine, namely that it was thinking like a human.

This imitation game later became known as the 'Turing test', and is accepted by many artificial intelligence researchers as the criterion of success. Indeed, the New York City philanthropist Hugh Loebner has offered $100,000 to any computer researcher who manages to pass it.[8] Turing's first objective in proposing it as a test, however, was less to prove the technology than that intelligence is an independent quality of the way symbols are manipulated – the way language is used. We see the thinking somewhere in the words, in what they say to us. This means that anything that can manipulate symbols in the same way humans do is capable of displaying intelligence. This made the issue of whether or not the mind was a part of the mechanical universe, a process that ran on the great universal computer, a question about whether natural language is computable. Answering this question has been a preoccupation of computer science since Turing posed it. At least since Ada Lovelace wrote her comments on Charles Babbage's designs for his 'analytical machine', an explicit link had been made between symbol manipulation and computation: the two were seen as equivalent, with the most familiar computation being the form that manipulated mathematical symbols.

Naom Chomsky's book *Syntactic Structures*[9] in many respects founded linguistics by attempting to provide a scientific basis for the study of language. Chomsky aimed to discover a 'universal grammar' common to all humans who use language by developing the concept of 'generative grammar', that is a 'machine', in the abstract Turing sense, that can be used to generate all the grammatical sentences in a given language. By specifying this machine, Chomsky aimed to provide a formal method for specifying what can be said in that language, just as the Turing machine was, in a sense, used to specify what was 'sayable' in the language of mathematics.

Chomsky specified three different types of generative grammar machines. Of these, the 'transformational grammar', was the most significant because it seemed to provide a specification for a machine that was not simply specifying the *syntax* of a language, the allowable ways that words, regardless of their meaning, could be arranged, but about specifying the *semantics*, the way that meanings are generated. 'Using his transformational rules', wrote the mathematician John Casti, 'Chomsky was able to show that ambiguous sentences such as "I like her cooking" could be given a single surface structure from several deep structures, while semantically equivalent sentences of the sort involving just a change from active to passive voice could have different surface structures emerging from the same deep structure.'[10]

Several computer researchers have attempted to use such grammars to translate languages. A team led by Robert C. Berwick at MIT have, for example, attempted to develop a 'universal parsing machine', a machine that can isolate the simple syntactical structures that are common to different languages.[11] Berwick succeeded in using his program to reveal grammatically equivalent sentences in English, German, Japanese and the Australian aboriginal language Warlpiri (of interest to linguists because words can be used in virtually any order). But his program could not reveal semantic equivalence – in other words, translate. The question remains whether a language's syntax being computable proves that its semantics, its meaning, is as well.

Computationalists never seem to have doubted that it was just a matter of time and technology before language would yield to the mathematical model. During the 1960s and 1970s, a number of research projects were doggedly undertaken to produce candidates for the Turing test. None needed testing. 'Talking computers' were no competition, no better at talking than robots turned out to be at walking. Their linguistic awkwardness was summed up by a poem entitled 'ARTHUR takes a test for divergent thinking' broadcast on BBC radio in 1974. ARTHUR, the computer, was asked by a Turing test interrogator to enumerate the uses of a paper clip:

> It can clip paper,
> It can clip papery substances.
> It can clip sheets.
> It can clip leaves. Can it clip leaves?
> *Yes, Arthur, it can.*
> It can clip branches.
> *No, Arthur, not branches.*
> It can not clip branches . . .
> Can it clip hedges?
> *No, Arthur, not hedges.*
> It can not not clip hedges.

ARTHUR was an acronym, naturally, for Automatic Record Tabulator and Heuristically Unreliable Reasoner, and his words were not the product of any mechanical procedure but of the imagination of the poet Laurence Lerner, seeking to show the vanity of any human attempt to make a machine think. What ARTHUR lacked, of course, was not the ability to manipulate words; he lacked the ability to understand them.

Computationalists have assumed that language could, like mathematics, be purified. Just as the foundations of mathematics were shown to be independent of physical reality, so the foundations of language, the universal grammar, could be, too. It is, from a computational viewpoint, an abstract system for manipulating symbols, with meaning coming from its application to real experiences – the study of semantics, in other words, is just 'applied' linguistics. The problem with this view is there is very little evidence to support such an assumption. 'Natural' languages do not have the qualities of purely formal systems, they are not just a calculus. The 'artificial intelligentsia' gained encouragement for their position from the successful development of programming languages like FORTRAN, COBOL and BASIC. But programming languages are formal languages; they are designed to be computable. The use of recognizable words such as PRINT and LIST does not mean that the computer understands what the words 'print' and 'list' mean, these words are merely used as mnemonics for the benefit of the programmer. They could arbitrarily be changed to any other symbol – WEEP and CONFIDE, say – and still have the same role, the same 'meaning'.

It is but a small step from the most superficial discussion of the issues surrounding linguistics to the deepest philosophical issues, but one we shall, at least for the time being, deny ourselves because, by the 1970s, artificial intelligence was far more concerned with practicalities than with principles. The funders of AI research, notably the military, were beginning to demand results. In Britain, the Science Research Council, which allocates academic research funds, commissioned a mathematician, James Lighthill, to assess the field's prospects. He reported that laboratory results could never be reproduced in the 'real world', because the real world is simply too complex to reduce to a formal grammar. The result was that the British programme came to an almost complete halt.

In America, meanwhile, artificial intelligence researchers were quickly learning to adapt to a new climate. They hurriedly downgraded their expectations, and began to concentrate on trying to find uses for their research in those 'real world' situations that could be simplified. What they discovered was the 'expert system', a computer that could perform the job of an expert. Expertise might sound like an even more ambitious aim than intelligence, in that the latter is usually regarded as a prerequisite of the former. However, experts from car mechanics to corporate lawyers usually work with well-specified sets of rules and techniques. If these rules and techniques could be codified, then perhaps the computer could reproduce some of the expertise, thereby automating increasingly expensive activities.

This shift of emphasis was reflected elsewhere in computer research. Attempts were made to reproduce human senses, the ability to recognize patterns and objects, and to identify sounds. In military research, the word 'intelligence' was substituted with the word 'smart', a word that, at least within computing circles, is far more easily applied to a machine. 'Smart' weapons were the direct result of research into pattern-recognition systems and the development of quite separate technologies, such as new types of radar. They were 'smart' only insofar as they could recognize a target and direct themselves towards it. They were not 'intelligent' in any challenging meaning of the word.

In the 1980s, full-blown ambitions to achieve artificial intelligence resurfaced in what was promoted as a sort of substitute space race between America and Japan. The objective: to develop the first thinking 'fifth generation' computer. To achieve this the Japanese government instituted an entire civil research programme under the auspices of the Institute for New Generation Computer Technology (known as ICOT for short – an acronym that seemed somewhat cheekily to echo Disney's EPCOT, 'experimental prototype community of tomorrow'). The Americans, meanwhile, instituted the Strategic Computing Program under the Defense Advanced Research Projects Agency, DARPA, the agency which was originally set up in 1958 following the national humiliation of the Russians putting the first man in space. The competition between these two was probably as much a device to reinforce political wills and budgetary commitments as an attempt to be the first to reach a particular technological target. In any case, neither was able to take the giant leap for mankind. The Japanese project did not produce, as they say in football, a result. By the time it was coming to the end of its ten-year term, researchers were downplaying the idea of achieving a single breakthrough in favour of the justification that it produced a lot of interesting technological innovations along the way – the very same justification used when various members of the US congress wondered aloud what the benefits of the NASA space programme were.

The Americans were never in competition in any case – or perhaps they were running on a different racetrack. Japan immediately embarked on what was informally dubbed the 'sixth' generation project (a clear case of counting your generations before they have hatched), which was to last ten years cost ¥200 billion, and develop computers of a yet higher order of complexity.

Reflecting the nationalistic element of these projects, Europe attempted to produce a response as part of its ESPRIT research project, and even Britain tried to revive its one-time central role in the field with the Alvey programme. None of these attempts have resulted in a machine that can be claimed to be capable of thought. Certainly none have provided a serious candidate for the Turing test. The question is whether that is a result of some fundamental flaw in the whole concept of artificial intelligence, or simply a failure of the technology.

Turing himself predicted that by the year 2000 there would be a machine with a storage capacity of one billions bits that would have a 70 per cent chance of passing a five-minute long Turing test conducted by an 'average' interrogator. He is already wrong in terms of the storage capacity, underestimating by several orders of magnitude the storage capacity of future computers (he was writing, it should be recalled, before the age of silicon chips). But at least this error seems to work in his favour, especially since AI researchers have, in their increasingly desperate search for new approaches to an intractable problem, started to consider the brain, rather than the mind and language, as the key to reproducing intelligence. If the brain's complexity could be matched, perhaps so too would its capacity to think. And there is a very serious possibility that computers will one day match the processing power and storage capacity of the human brain.

Hans Moravec, Director of the Mobile Robot Laboratory at Carnegie Mellon University in Pittsburgh, is engagingly optimistic about the future capabilities of computer technology. In his book *Mind Children*,[12] he pursues a train – more a rollercoaster – of thought through what he sees as the unfolding story of robotics and AI. Part of his approach is to draw graphs, all emphatically pointed in an upward direction, comparing the specification of different machines, both real and imagined, and animal species. One graph, labelled 'Comparative computational power and memory', compares bacterial reproduction (processing power, about 10^3 bits per second; capacity, about 10^6 bits), the Cray 2 supercomputer (10^{12}, 10^{12}), the American public telephone system (10^{13}, 10^9) with a human being (10^{14}, 10^{14}).[13] Another graph, entitled 'A century of computing' shows the ineluctable growth of computing power beginning with clockwork calculators such as the analytical engine designed by Charles Babbage (which Moravec dates to 1910, when he reckons it would have been built had Babbage and his successors persisted with the project), which he rates as having a computational power related to cost of little more than manual calculation to the Cray 3 (a hypothetical machine at the time he wrote the book) which he rates at around 1,000 bits per second per dollar (at 1988 values). From there he draws a straight line to around 10^7 bits per second per dollar, achievable by the year 2010, which would mean the capability of producing a supercomputer with the same amount of computational power as a human, and on up to around 10^{10} bits per second per dollar by 2030, when it would be possible to built a personal computer with the computational power of a human at a cost of a few thousand dollars.

Moravec is a refreshing character in an industry inclined to hyperbole. His wildly extravagant estimations are backed up by pretty solid assumptions and calculations (his book contains a complete appendix on his method of measuring computer power) and a mind unconstrained by the convention of scientific conservatism. He is what one might call a strong computationalist; he sees no limit to the reach of computation, nor to the power of computers. He also uses his hyperactive mind to perform some pretty weird thought experiments, including one designed to show how, when the universe cools down, as cosmologists predict it will, so the energy needed to perform a computation will fall. Moravec outlines a plan for dealing with this eventuality: 'Before it's too late (better hurry, there are only some trillions of years left!) we take some of the remaining organized energy in the universe and store it in a kind of battery.' Moravec proposes a beam of photons pushing apart two mirrors suspended in space. 'The idea is to use about half the energy in the battery to do T amount of thinking, then to wait until the universe is cold enough to permit the *remaining* energy to support another T, and so on indefinitely.' It's crazy, but it might just work.

So, Moravec hopes, will the assumption that sufficient computational power will produce intelligence. One of the developments that reassured him was the phenomenon of 'emergence'. As clearly demonstrated by the Game of Life – indeed by conventional board games – the simplest systems can, given time and space, produce quite unexpected and surprisingly complex results. Perhaps thinking is an 'emergent' property of a very complex machine based on very simple principles. What a wonderful idea. Find the principles, develop a sufficiently complex machine, and out will pop thought.

Danny Hillis set up a company to develop a 'parallel' computer he had designed for his doctorate at the Massachusetts Institute of Technology. He called his computer the 'Connection Machine'. He called the company, in a clear statement of his ambitions, Thinking Machines. The Connection Machine was an awesome system. Unlike conventional computers, which have a single processor that handles each instruction sequentially, one after the other, a parallel computer has many processors, in the case of the Connection Machine – a huge, sinister-looking black cabinet covered in winking red lights – 65,536 of them.

Designing a parallel computer is not quite as easy as it might seem, because in order to be of more value than a lot of single-processor or 'serial' computers strapped together, there has to be some method of allowing the processors to communicate with each other, to 'interact'. This Hillis achieved by creating the equivalent of a telephone network through which the processors would send each other messages. This network was a beautiful abstract design – a configuration that took the form of a 12-dimensional cube. The result was a machine with immense raw computing power, but also a power to model 'real-world' phenomena. It could allocate a processor to each element of the phenomenon to be modelled – say, one processor per particle in a swirl of smoke – and use the network to reproduce their interactions.

The Connection Machine looked like a very promising candidate for the sort of machine from which interesting phenomena might emerge. Hillis himself explored this possibility in a paper published in 1988 entitled 'Intelligence as an emergent behaviour; or, the songs of Eden'.[14] 'It would be very convenient' if intelligence were an emergent behaviour of randomly connected neurons,' he wrote. 'It might then be possible to build a thinking machine by simply hooking together a sufficiently large network of artificial neurons. The notion of emergence would suggest that such a network, once it reached some critical mass, would spontaneously begin to think.'

Hillis is quite sanguine about the concept of emergence. 'Emergence offers a way to believe in physical causality while simultaneously maintaining the impossibility of a reductionist explanation of thought. For those who fear mechanistic explanations of the human mind, our ignorance of how local interactions produce emergent behaviour offers a reassuring fog in which to hide the soul.' The ghost is reintroduced into the machine. Hillis nevertheless is a keen proponent of AI, seeing emergence as providing a convenient new line of research into the subject.

The emergence theory of intelligence – if theory is not too grand a word for it – is part of what has been called the 'connectionist' movement in AI. 'Connectionism' is yet further proof of the metaphorical allure of prevailing technology. In the 1960s and 1970s, researchers hoped to find intelligence in the computer. In the 1980s, the era of the global telephone system and the 'teleworker', the secret lay in the network. The mind lay in the brain's wiring, a property of the brain that could be easily reproduced in a computer by creating what came to be called a 'neural net', a network of simulated neurons that could interact with each other in the way biological neurons interact.

Throughout its short history, artificial intelligence research has engaged in two quite different enterprises: attempting to use computers to reproduce intelligence behaviour and trying to discover whether the mind *is* a computer. The former is sometimes known as 'weak' AI, the latter as 'strong' AI. Connectionism promises much advancement in the former, especially if you accept weak AI's increasingly modest threshold of intelligence: the ability to distinguish a washer from a bolt, for example. But it has nothing to do with the latter. Nor, indeed, do any existing AI systems; they are no more intelligent than animated cartoons are alive.

In 1965, the philosopher Hubert Dreyfus wrote a paper for the Rand Corporation called 'Alchemy and artificial intelligence' in which he attacked the increasing enthusiasm among AI researchers for developing programs that played games such as chess. This was, Dreyfus argued, no more evidence that the computer was approaching a state of intelligence than a monkey climbing a tree was nearer reaching the moon. In what came to be known as *L'Affaire Dreyfus*, he rashly took up a challenge to play a computer at chess, and lost, a result that the AI community gleefully reported under the heading 'A ten year old can beat the machine – Dreyfus' with the subheading 'But the machine can beat Dreyfus'.[15]

By the end of the 1980s, computers were competing with grand masters, apparently proving that at least one form of intelligence could be artificially reproduced using a computer. But who is to say that playing chess is proof of intelligence? Why is not the ability of mechanically reproducing 'intelligent' chess moves a reflection on the nature of chess as much as the capabilities of the computer? Perhaps, as Dreyfus suggested following his defeat,[16] there are two ways of playing the game, mechanically or inspirationally. Perhaps a chess game is not the display of intelligence that chess enthusiasts assume it to be. For Dreyfus, the point remained that the attempt to create a 'disembodied' intelligence, a pure symbol-manipulating language machine, was doomed because intelligence is intimately connected with the human system that emodies it.

Much is made of the distinction between 'strong' and 'weak' artificial intelligence, but one has to ask whether they are, as the terms imply, really at two ends of a single scale. Weak AI states that computers can mimic intelligent behaviour. But so can a speak-your-weight machine. So what would be different about using a computer to do the job? The difference lies in whether computers can *simulate*, rather than just *imitate*, intelligence, which depends on whether the processes that produce the mind or intelligence are themselves mechanical (which means computational), and that is the very issue addressed by strong AI.

In his quite unexpected assault on artificial intelligence, Roger Penrose did not even bother with weak AI. He concentrated on developing a robust argument aimed exclusively at the strong variety. Penrose is Rouse Ball professor of mathematics at Oxford University and Stephen Hawking's one-time collaborator. Though a mathematician, Penrose is not impressed by big numbers the way Hans Moravec is. In his book *The Emperor's New Mind*, published in 1989, he imagined the building of a computer called the 'Ultronic' with 10^{17} logic units. 'That's more than the number of neurons in the combined brains of everyone in the entire country!', boasted the Ultronic's designer.[17] Even this, Penrose hypothesized, could never answer a simple question about its own feelings.

But Penrose's scepticism about AI is not that it is overly impressed with big numbers. It is the computationalist view that is his target. 'The belief seems to be widespread,' he wrote, 'that . . . "everything is a digital computer". It is my contention . . . to try to show why, and perhaps how, this need *not* be the case.'[18]

* * *

Penrose's office at the Oxford University Mathematical Institute is chaos: every horizontal surface is covered in a clutter of papers and books, every vertical surface is covered in mathematical graffiti. A Hollywood set designer could not have done a better job at recreating what most of us would imagine a mathematician's office to be like. Except there is no computer. This is not, Penrose claims, a sign of repressed technophobia, but a reflection of his belief that there is more to mathematics than computation.

One piece of evidence he supplies for this cheekily uses what has in many respects become the symbol of computationalism: the Mandelbrot set. The procedure used to generate the set is simple. You take a number, square it, square the result and add the number you started with, square *that* result and add the number you started with and so on. Performing this procedure with different starting numbers, Mandelbrot discovered that some numbers produced a series of new values that spiralled on up to infinity, while other numbers would settle down to a single result. The numbers that do not spiral off to infinity are the ones that fall within the set. In order to examine if there was any interesting feature common to the numbers that fall within the set, Mandelbrot plotted them on a graph, and so discovered that the characteristic warty potato shape that was to be honoured with his name. Various refinements were added, such as colour, where different colours reflected how many 'iterations' (successive squaring and adding operations) it took for the number to settle down to a fixed result.

Given a starting number, is there any way of telling whether it belongs to the set? Obviously, you cannot rely on waiting for a result, because it could only appear after a million, or a billion, or a trillion, or a 'googol' (one followed by a hundred zeros), or a googol googol, or a number equivalent to the number of particles in the universe times a googol times a googol iterations – or it may never appear, which you will only discover after having carried on iterating for ever. No, what you need is an algorithm that will predict whether or not the starting number is a member of the set, and, Penrose pointed out, at the time he was writing, no such algorithm existed.

The purpose of the demonstration was not to denigrate the Mandelbrot set, though there can be little doubt that he chose it precisely because of its association with computing enthusiasts. It was to show that there exists a sort of 'Platonic' realm (named after Plato) quite separate from the physical realm, where there exists objects such as the Mandelbrot set. These objects are not *invented* by mathematicians, they are *discovered*. For example, the number pi (the ratio of a circle's circumference to its diameter) has an infinite number of decimal places, like a recurring number, but with no particular pattern, so you cannot predict what number will be at a particular decimal place without working the number out to that level of accuracy. Suppose, then, no one ever bothered to work out pi to a googol decimal places. Is the proposition that the number at that decimal place is, say, '7' true or false? If you are, as they say, an 'intuitionist' and believe that numbers are invented as mathematicians carry out their calculations, and therefore do not exist independently of human knowledge of them, then the answer is neither. Penrose, however, is a Platonist. He believes that mathematical objects have independent existence, and the Mandelbrot set is a striking example of this. Mandelbrot did not draw it in the way that an artist might draw a picture of a warty potato. But neither is it demonstrably a computable object – though the computer allows us to glimpse it to ever finer degrees of detail. Platonic reality, in other words, cannot be explored by computer alone; to truly discover its secrets, you need insight.

Penrose's proof of this relies on Gödel's finding that arithmetic (or any 'formal system') must contain a proposition that cannot be arithmetically proved. Such a proposition has to be true, even though the system that produced it cannot prove it to be true. It is this truth which, Penrose argues, can only be seen from without the system, with the benefit of human insight (or perhaps 'outsight'). 'Mathematical truth is something that goes beyond mere formalism', wrote Penrose. 'This is perhaps clear even without Gödel's theorem. For how are we to decide what axioms or rules of procedure to adopt in any case when trying to set up a formal system? Our guide in deciding on the rules to adopt must always be our intuitive understanding of what is "self-evidently true", given the "meanings" of the symbols of the system.'[19]

Penrose is not quite a mathematical maverick, but he is certainly independently minded, and inherited from his father Lionel, a respected physician, an enthusiasm for mathematical invention – indeed, for all forms of invention. Together, he and his father tried to simulate self-reproduction, complete with a crude form of genetics, using home-made wooden model 'molecules'. In 1958, they published drawings of an 'impossible staircase' and 'impossible triangle', optical illusions popularized by the artist Escher, whose images were to become icons of the logical circularities and paradoxes identified by Gödel, which Penrose cited as evidence for the need for mathematical insight. He also designed sets of tiles which could cover a surface of any size in a pattern that at no point repeats itself – and even contemplated marketing them as a floor covering.

There is nothing in particular that links these ideas, except that they are beautiful examples of what 'insight' can do. Each one takes an unexpected, tangential approach to revealing a heavily concealed problem. The impossible staircase is a classic example. In one of Escher's portrayals of the idea, entitled 'Ascending and descending', we see lines of robotic creatures marching up and down a perpetual staircase arranged in a rectangle, with the top step of each flight becoming the bottom of the next. It is a perspectival trick, a mere illusion, yet one which feels like genuine entrapment into the closed loop of formal systems, systems which seem hermetically wrapped up in their self-contained, self-referential worlds. Yet, as Penrose showed in his drawings, we can *see* that they are stuck in a loop.

The idea that Gödel's theorem in some way proves the impossibility of mechanical minds is not new. The British philosopher John Lucas introduced the idea in his 1961 paper 'Minds, machines, and Gödel'.[20] 'At one's first and simplest attempts to philosophize,' Lucas wrote, 'one becomes entangled in questions of whether when one knows something one knows that one knows it, and what, when one is thinking of oneself, is being thought about, and what is doing the thinking'. An entanglement indeed, one which can be expressed in the form of ditty:

> Often I have sat and thought,
> Philosophically,
> Am I the author of that thought
> Or is it the author of me?

Who knows? Which is, of course, the point. 'The paradoxes of consciousness arise because a conscious being can be aware of itself, as well as of other things, and yet cannot really be constructed as divisible into parts', wrote Lucas. 'It means that a conscious being can deal with Gödelian questions in a way that a machine cannot, because a conscious being can both consider itself and its performance and yet not be other than that which did the performance.'

Several elegant attempts to refute Lucas's argument were outlined by Douglas Hofstadter in his classic exploration of the nature of self-reference and strange loops, *Gödel, Escher, Bach: the eternal golden braid*. Hofstadter had a spell writing the *Scientific American* column written previously by Martin Gardner and subsequently by A. K. Dewdney. And, like the others, he is a keen member of what might be called the Penrose school of mathematics, delighting in the exercise of mathematical insight, in the playing of the game.

Hofstadter is, above all, the master thought experimenter. There is no argument that is immune from being restated in terms of a hypothetical event, usually taking the form of a mock Platonic dialogue between characters with punning names and an over emphatic mode of expression. Such a character is 'Loocus the Thinker'. One day Loocus, who has led a sheltered life, encounters a woman. 'Such a thing he has never seen before, and at first he is wondrous thrilled at her likeness to himself; but then, slightly scared of her as well, he cries to all the men about him, "Behold! I can look upon her face, which is something *she* cannot do – therefore women can never be like me!"' Not to be so easily dismissed, the woman smartly replies: 'Yes, you can see my face, which is something I can't do – but I can see *your* face, which is something *you* can't do! We're even.'[21] Loocus, in other words, overestimates the power of the human mind to perform the 'paradox of consciousness', to be aware of itself.

The artificial intelligence debate is a bit like Penrose's impossible staircase: a constant uphill struggle that gets you nowhere. Its importance here is whether or not the computationalist position holds, whether or not the universe really is a computer. Penrose's answer to that question is an emphatic 'no'. In this he is definitely swimming against a strong tide of opinion.

* * *

In his intriguing paper 'Theatricality and technology', Julian Hilton of East Anglia University describes theatre as a 'complex aesthetic machine, dedicated to the representation of the imaginable through performance'.[22] Furthermore, it is a machine that, Hilton argues, could be used to reproduce intelligence. Just as it enables actors to play characters that are more intelligent than themselves, a similar sort of representational mechanism could be used to reproduce intelligence in a computer.

It is an intriguing suggestion, with much to recommend it. However, what such a mechanism will succeed in doing is not *simulate* intelligence but *imitate* it. Simulation is only possible where there is a mathematical model, a virtual machine, representing the system being simulated. Theatre is not a machine, at least, it has not been proven to be so – which is why it is such a powerful expressive form.

The obsessive – and, I would argue, fruitless – quest to reproduce human consciousness and intelligence is not a sensible computationalist target. Consciousness and intelligence cannot be precisely defined, let alone explained – and certainly not measured, as supporters of the IQ test assume it can be. However, this is no more a catastrophic limitation to the computationalist dream than is the inability of computers to model beauty or charm. Such qualities are simply not scientific properties or mathematical parameters. The study of non-linear systems has shown that the most natural, apparently non-mechanistic phenomena – from hurricanes to evolution – can be computed. As the excitement surrounding chaos theory hinted, even social and cultural phenomena, may, at least on the level of statistical generalizations, reveal computable structures. Cellular automata have shown that life itself may turn out to be computable.

Even if one accepts the role of humans as a unique case, the observer that is necessary in any closed system to save it from Gödel's theorem, and even if one accepts Penrose's belief that there are natural phenomena that can be mathematically described without being computable (ultimately, what he hopes will prove the case with human intelligence), even if one accepts these things, most of the world we live will still yield, in principle if not in practice, to simulation. Indeed, the physical realm could be regarded as a simulation of a deeper, purer reality – a virtual reality. Virtual realists certainly believe it to be so. For them, Ivan Sutherland's head-mounted display is more than a glorified VDU. It is the portal into a new, unexplored world; into cyberspace.

Notes

1 René Descartes, *Discourse on Method and the Meditations*, trans. F. E. Sutcliffe, London: Penguin, 1968.

2 Ray Monk, *Ludwig Wittgenstein: the duty of genius*, London: Cape, 1990.

3 Gilbert Ryle, *The Concept of Mind*, London: Penguin, 1963.

4 See Derrik Mercer (ed.) *Chronicle of the 20th Century*, London: Chronicle/Longman, 1988, p. 639.

5 Hodges, 1985, p. 290.

6 Hodges, 1985, p. 291.

7 Alan Turing, *'Computing machinery and intelligence'*, *Mind*, reprinted in Douglas Hofstadter and Daniel Dennett (eds) *The Mind's I*, Brighton: Harvester, 1981.

8 Mark Tran, 'A testing time for "human" computer', *Guardian*, 8 November 1991, p. 1.

9 Naom Chomsky, *Syntactic Structures*, The Hague: Mouton, 1957.

10 John L. Casti, *Paradigms Lost*, London: Scribners, 1989, p. 230.

11 John Horgan, 'Word games', *Scientific American* 265 (4) October 1991, p. 19.

12 Hans Moravec, *Mind Children: the future of robot and human intelligence*, Cambridge, Mass.: Harvard University Press, 1988.

13 Moravec, 1988, p. 61.

14 W. Daniel Hillis, 'Intelligence as an emergent behaviour: or, the songs of Eden', *Dædalus: Proceedings of the American Academy of Arts and Sciences*, 117 (1) Winter 1988.

15 In Sherry Turkle, *The Second Self: computers and the human spirit*, New York: Simon & Schuster, 1985, p. 240.

16 See Hubert Dreyfus, *What Computers Can't Do*, New York: Harper & Row, 1979.

17 Roger Penrose, *The Emperor's New Mind: concerning computers, minds, and the laws of physics*, Oxford: Oxford University Press, 1989, p. 1.

18 Penrose, 1989, p. 23.

19 Penrose, 1989, p. 111.

20 J. R. Lucas, 'Minds, machines, and Gödel', *Philosophy* 36, 1961.

21 Douglas R. Hofstadter, *Gödel, Escher, Bach: the eternal golden braid*, London: Penguin, 1980, p. 477.

22 Julian Hilton, 'Theatricality and technology: Pygmalion and the myth of the intelligent machine', unpublished paper, 1988.

6
CYBERSPACE

'Cyberspace has a nice buzz to it,' said William Gibson, recalling his use of the term in his fiction, 'it's something that an advertising man might of thought up, and when I got it I knew that it was slick and essentially hollow and that I'd have to fill it up with meaning.'[1] By the beginning of the 1990s it was so full of meaning, it was fit to burst. In his original use of the term in *Neuromancer*, Gibson had famously described it as a 'consensual hallucination' – a term he left ill defined. It was, he later said, meant to suggest 'the point at which media [flow] together and surround us. It's the ultimate extension of the exclusion of daily life. With cyberspace as I describe it you can literally wrap yourself in media and not have to see what's really going on around you.'

It was a fictional rendition of Ivan Sutherland's original concept of the 'ultimate display', a form of display that presented information to all the senses in a form of total immersion. However, Gibson had extended Sutherland's idea of a 'looking glass into a mathematical wonderland' to embrace the entire universe of information: 'A graphic representation of data abstracted from the banks of every computer in the human system. Unthinkable complexity. Lines of light ranged in the nonspace of the mind, clusters and constellations of data. Like city lights receding.'[2]

In the rhetoric of the virtual realists, this 'nonspace' was not simply a mathematical space nor a fictional metaphor but a new frontier, a very real one that was open to exploration and, ultimately, settlement. 'Cyberspace . . . is presently inhabited almost exclusively by mountain men, desperados and vigilantes, kind of a rough bunch', said John Perry Barlow. 'And, as long as that's the case, it's gonna be the Law of the Wild in there. . . . Whenever you make a financial transaction, really it involves electronic data representing money. So we feel that the way to minimize anxiety, and to make certain that the freedoms we have in the so-called real world stay intact in the virtual world, is to make it inhabitable by ordinary settlers. You know, move the homesteaders in.'[3] To do this, Barlow had even set up the Electronic Frontier Foundation with Mitch Kapor, the founder of the software house Lotus, one of the most successful companies to emerge out of the personal computer era and one, ironically, roundly condemned by the hacker radical Richard Stallman for its illiberal policy towards copyright.

In the twilight of the space age, cyberspace was becoming the new final frontier, and virtual reality was the *Enterprise*. NASA's key role in the development of the technology, right at the time of the *Challenger* and *Hubble* humiliations, when the agency no longer commanded the emblematic heights (or generous financial backing) it had enjoyed during the 1960s, was heavily symbolic. Like astronomical space, cyberspace was only dimly perceived by ordinary people, but there was a promise that technology would one day provide them with access to it. Day-trips to the moon having proved unfeasible, attention was turning to mystery tours of the digital domain.

What, then, is cyberspace? The actual term is technically unimportant. Other phrases are used synonymously: cyberia, virtual space, virtual worlds, dataspace, the digital domain, the electronic realm, the information sphere. One can examine its etymological entrails for meaning – 'cyber', meaning steersman, coming from 'cybernetics', the study of control mechanisms – but that yields very little. A more productive strategy is to try to discover why the terms have acquired such currency. How did a word that Gibson had thrown into his work almost casually when he coined it acquire, within a few years, such value?

One interpretation of cyberspace is that it concerns the annihilation of space. 'As electrically contracted,' wrote Marshall McLuhan in 1964, 'the globe is no more than a village.'[4] 'After three thousand years of explosion, by means of fragmentary and mechanical technologies, the Western world is imploding. During the mechanical ages we had extended our bodies in space. Today, after more than a century of electric technology, we have extended our central nervous system itself in a global embrace, abolishing both space and time as far as our planet is concerned.'

Neglected in the 1970s, when the world was probably too diverted by shortages of energy to concern itself with ideas of electrical contraction, this concept of the global village began to become fashionable again in the 1980s. It was seen as the perfect expression of the new era of world finance and international telephone networks. In the 1980s, the financial system had migrated onto computer and communications networks, satellite and cable links that spanned the globe, capable of carrying data and voice, creating the conditions that produced the October 1987 stock market crash, where a fall in the New York stock exchange precipitated a collapse in prices that tripped off markets around the world within hours. 'In principle,' wrote Mark Poster, a Californian professor of history, in 1990, 'information is now instantly available all over the globe and may be stored and retrieved as long as electricity is available. Time and space no longer restrict the exchange of information. McLuhan's "global village" is technically feasible.'[5]

The imagery of the 'global village' is seductive, suggesting that the technology of communications will collapse dispersed urban alienation into the cosy confines of a pre-industrial age. It suggests the emergence of a whole new type of working environment, the 'telecottage', set in rural surroundings, far away from the jams yet part of the world's information traffic. Telecommunications companies have even started setting up such cottages in remote areas such as the Scottish Highlands. Certain white collar-jobs, claim the global villagers, entail nothing more than the exchange of information – meetings, paperwork, taking decisions – all of which can easily be communicated over the telephone network. As a result, a teleworker's attachment to the job becomes a factor of his or her connection to the network rather than proximity to the company office.

But Marshall McLuhan expected more from the idea of the global village than a new type of post-industrial working environment. He saw technology as an extension of the body. Just as the wheel is an extension of the foot, the telescope an extension of the eye, so the communications network is an extension of the nervous system. So, as the communications network has spread across the globe, so has our neural network. Television has become our eyes, the telephone our mouths and ears; our brains are the interchange for a nervous system that stretches across the whole world – we have breached the terminating barrier of the skin.

The technology that has made this possible is the network. Networks are not new: there have, presumably, been social ones since the dawn of society. What is new is the technology of communication that has enabled information of any type to be carried from one place to another regardless of their distance. Such networks are electronic, and carry their messages instantly by wire, by optical fibre, by radio and microwave. In computing, it is the network that is seen as providing the next great step in the technology's inevitable progress. Personal computing will become what Steve Jobs has dubbed 'interpersonal computing'.

Networks first found their way into computing via the development of a system to connect computers deployed in projects funded by the American Advanced Research Projects Agency, ARPA. This ARPAnet was originally designed to allow APRA researchers to share data, but was increasingly used to exchange messages, which in turn helped develop a sense of community between the geographically scattered centres they worked in. This was, in a sense, the world's first 'virtual' community, existing only in its interaction over ARPAnet.

ARPAnet developed in two directions: it grew outwards, to span the entire globe, coming to be called Internet. Internet is, for many, the model of what virtual communities can be like. Geography is irrelevant – so there are no 'backwaters', no 'provinces' excluded by the central metropolis. Hierarchy is irrelevant, because everyone has equal access to the network, and everyone is free to communicate with as few or as many people as they like.

ARPAnet also contracted, inspiring the development of the world's first 'local area' network, 'Ethernet'. As the word 'local' implies, geography has not quite been rendered irrelevant by telecommunication networks. The more information there is to carry and the further it has to go, the more it costs to carry it. Local area networks (LANs) were designed to carry a great deal of information across short distances – the distance between computers in the same office. Because of their carrying capacity –

their 'bandwidth' – they are able to create a much richer form of communication between the individual users, so rich, in fact, that on well-implemented local area networks, the distinction between each user's personal system and the network begins to blur. What belongs to whom in terms of files and facilities depends not on its physical location inside any particular machine, but upon the way it is organized across the LAN. In 1991, NeXT, the computer company set up by Steven Jobs after his departure from Apple, announced 'Zilla', a program that enables a network of NeXT computers to act as a single, virtual supercomputer. A network of 100 NeXT machines was, according to Zilla's designer Richard Crandall, as powerful as a Cray 2, then the most powerful supercomputer in the world. A Zilla computer could, furthermore, have its power distributed between offices across the globe. With the development of optical fibre, which offers almost limitless bandwidth, as an alternative to wire, the possibility arose that this new form of 'interpersonal' computing would truly render geography irrelevant and allow new forms of social as well as commercial interaction to emerge.

In the mid 1980s, NASA's Human Factors Research Division began working on developing what it provocatively called 'telepresence'. Telepresence was originally promoted as a way of controlling robots. Since it is better to send machines than humans into hazardous environments like space, the NASA research team aimed to provide a wrap-around technology that would give the machine operator the feeling of 'being' in the place of the machine being operated. If, for example, a robot was being used to repair an external component on a shuttle or space station, the robot's cameras would be connected to a head-mounted display so that the wearer could see what he or she would see if actually there. Similarly, a glove or exoskeleton (that is, a series of hinges and struts clamped to the body, like an articulated splint) could be used to reproduce the wearer's movements in the robot's arm and hand or to provide tactile feedback, resisting the wearer's movements in accordance with the pressure exerted on the robot's limbs when it picks up an object. A sufficiently rich communication between the operator and the robot would, the NASA team hoped, result in the robot essentially becoming the operator's body; he or she would be 'telepresent', instantly transported to wherever the robot was working.

Since the communication between robot and operator is via an information link, perhaps you could be telepresent anywhere. Just plug the helmet and some sort of sensory bodysock into the telephone and television network and you would be away, literally, discovering the true meaning of McLuhan's limitless sensorium. No more watching television, but, as William Gibson put it, 'doing' television. We could climb through the window of the world.

Is this, then, cyberspace? Is there really some sort of space, some sort of independent realm created by the interconnection of the world's information systems? Is this a metaphorical space or a real one?

* * *

Conventionally, we think of networks as transparent, as systems for conveying messages from one human being to another without in any way shaping their meaning. McLuhan's other great contribution to the media age was the phrase: 'The medium is the message.' What he meant was that networks (or media – in other words, systems for carrying information) are not transparent. Television is not a window on the world, it does not simply show its audience pictures of events that happen to be taking place elsewhere. Rather, it actually has a role in determining what the audiences see and how they make sense of it. A game show or soap opera is not a natural event; it has been created specifically for the cameras, and only makes sense if seen on TV. (Studio audiences, one might think, are watching the event 'for real', but in fact they need the prompting of an army of studio managers and warm-up artists to make up for what is, in fact, a very unreal event, constantly interrupted by retakes, filmed inserts and cutaways.)

On 20 July, 1989, Wes Thomas, publicist for the magazine *Mondo 2000*, claimed that he had 'unleashed the world's first media virus'.[6] An unspecified source had told him of the spread of a particularly dangerous computer virus, variously called 'October 12', 'Datacrime' and 'Columbus Day'. A computer virus is extraordinarily like a biological one. A biological virus is a small strand of genetic code that uses the host organism's replicating mechanism to produce copies of itself, which are then spread to other host organisms via whatever medium they are able to use – say, the exchange of body fluids, as seems to be the case with HIV. A computer virus is similarly a strand of code, a computer program, that uses the computer's replicating mechanism to produce copies of itself, which are then spread to other host computers via whatever medium they are able to use, usually via floppy disks and the public networks used by enthusiasts to exchange software. The most significant difference between a computer virus and a biological one is that the former is written by a computer programmer for mischievous purposes, whereas the latter arises spontaneously in nature.

Thomas sent a note via electronic mail to two journalists alerting them to news of the computer virus, which he subsequently discovered would activate on 12 October in any year and wipe out the host computer's disk drives on the 13th, which in 1989 happened to fall on a Friday. As soon as the story appeared in the *San Francisco Examiner*, it was picked up by the world's press agencies. According to Thomas, the CNN cable news service (often described as the global village's local TV station) announced that 'all PC computers would be wiped out at 12.01am on October 13th'. The British tabloid, *Daily Star*, even featured the story on the front page of its 5 October issue, anticipating the spread of a 'doomsday computer bug', and the BBC carried a news item on 13 October on the anticipated epidemic, which, as it turned out, amounted to no more than an outbreak of irregular behaviour on a computer at the Royal National Institute for the Blind.

The interesting feature of the Friday the 13th virus was, of course, the way that the reporting of its spread was as much a feature of its epidemiology as the spread of the program itself. Thomas had released a story that was evidently highly contagious. Besides being a date of Christian significance (13 being the number at the Last Supper, and Friday the day of the Crucifixion), Friday the 13th had far more popular significance as the name of a successful series of Hollywood horror movies. The AIDS scare had also increased awareness of the threat of viruses. In the era of antibiotics and vaccines, the public perception of infection in countries like Britain and America had been that it was basically controllable, despite the continuous background presence of venereal disease. Few people realized just how little control medicine had over viral disease. AIDS changed all that. It also disinterred old attitudes about the links between disease and sin. The term 'virus' was no longer simply a medical term, like 'bacteria'; it had suddenly acquired a moral resonance. It was a resonance that was easily transferred to the computer virus. Like AIDS, we were all threatened with being tainted by the hacker's lack of moral hygiene. Infection was guilt.

The technological as well as the social climate provided the Friday the 13th virus with perfect breeding conditions. Since the international acceptance of the IBM Personal Computer as a technical standard, the vast majority of personal computers had become functionally identical: a homogeneous 'species' had, in other words, emerged with very little variation to protect it from opportunistic infection. Worse, this technological monoculture was both dispersed, used across the Western world by the self-employed and the multinational alike, and highly connected by computer networks and dial-up computer services such as bulletin boards. The release of a virus capable of efficient reproduction in such an environment genuinely threatened widespread damage.

But not *that* much damage. Because the story had such mythical potency, the technical threat of the computer virus was wildly overstated by the news coverage. The world portrayed as under the greatest threat, the world of the corporate computer user, was in fact under the least threat, because most corporations get most of their software from authorized channels. The most vulnerable users were those who regularly swapped illicit software – the hackers, in other words, who were best able to spot any signs of infection and deal with them, and who in many people's eyes would be the deserving victims of any damage they caused.

More significant than the impact of the Friday the 13th virus was the increasing awareness, expressed by writers like Thomas, that the international media network was, like the computer network, not merely a passive communications system but an environment that, in many respects, had a life of its own. The virus story was a global village phenomenon. AIDS, a global village disease spread by international travel and the exchange of blood products, had created the conditions for its spread. It was introduced into an environment which was increasingly homogenized by global capitalism, and that, thanks to the sheer speed of turnover of the news agenda, was unable to build resistance. It was, in other words, evidence of the emergence of an artificial environment – something that McLuhan himself anticipated.

According to Richard Dawkins, a zoologist at Oxford University and the author of *The Selfish Gene*,[7] a feature common to all forms of life is the replicator, a mechanism that can store information and replicate it. The replicator for natural life is the gene, made up of DNA. An important distinction needs to be established in order to understand the concept of the replicator. There is the information to be replicated, called the genotype, and the product of that information being 'expressed', the phenotype. The information encoded in the gene is the genotype. The organism it produces is the phenotype. The phenotype can be thought of as a set of tools which are there to make copies of the genotype. Our bodies, along with their reproductive apparatus, are the most obvious phenotypic expression of our genes. Dawkins called his book *The Selfish Gene* because, from a zoological perspective, living organisms are the product and servant of their genes. The genes that produce the most effective phenotypes for reproduction do best, because they spread. Those that do badly disappear.

Computer viruses are, like biological viruses, replicators in a special sense, because they are parasitical, they do not come packaged with the means of phenotypic expression. Nevertheless, it is easy to imagine a refinement of the computer virus that fits in exactly with the genotype/phenotype model. The virus program can be thought of as the genotype, and the effect it has on the computer when it is executed can be thought of as the phenotypic expression. The program might even be miscopied, and the result might be a new version of the original computer virus program that, when executed, produces a phenotype that copies the virus more efficiently. The result would be the beginnings of evolution.

The unfolding scenario is a threatening one: the artificial environment created by computers will become host to a new, evolving lifeform, one that could spread uncontrolled through the world's networks. We should not, however, reassure ourselves with the thought that such a lifeform will confine itself to the realm of computers. Let us extrapolate further by introducing one of the most powerful of Dawkins's ideas: that of the extended phenotype.

The phenotypic effect of a gene is usually thought of as the body that carries the gene. But why stop there? Surely, argues

Dawkins, anything that results from the gene's expression can be regarded as the phenotype.[8] Even something as large and inanimate as a beaver's dam is a genetic effect, one that contributes to the beaver gene's chances of reproduction. By the same token, could not a computer virus evolve some sort of extended phenotype? Suppose, for example, that a mutated virus caused the host system to behave in a way that prompted the computer user to indulge in some particularly vigorous file copying (perhaps it causes the system to crash a couple of times, which encourages the user to make more backup copies of infected files). Such a virus could spread further, and would do so because it, unlike other viruses, had evolved a mechanism for manipulating the world beyond the computer.

There is no reason to suppose that evolution would stop there. Human users provide a perfect means of acquiring new reproductive resources. They are influenced by the computers – otherwise they would not use them – and could therefore be used in any way that escapes obvious detection to aid a particular virus's spread. This does not mean that viruses would turn computer users into mindless servants, any more than biological viruses that use us, as they surely do, to spread, have turned us into mindless servants. The point is that replicators are almost by definition opportunistic. Successful ones will take any opportunity they can to spread; that is what made them successful and is the reason they, unlike all the unsuccessful ones, have survived and evolved.

Although there is very little evidence that such replicators have yet colonized our computer systems, on the same day that Wes Thomas's media virus swept the globe, an event occured that did seem to confirm that something mysterious was spreading through the global village.

* * *

On Friday, 13 October, 1989 Wall Street went into free fall, throwing the world's financial markets into a state of chaos which, for many dealers, chillingly recalled the dark days of October 1987. The strange aspect of the 1989 crash was, however, the total absence of any apparent cause: the world economy was in pretty good shape and the European, Far Eastern as well as US stock markets, until that point, appropriately stable. 'Explanations abound,' reported *Financial World*. 'An isolated sequence, driven by events perhaps? An illogical fall with a fast return to reason? Or, a chilling thought, the reverse?'[9]

There was no possibility that the trigger was the computer virus. Computers had been implicated in 1987 crash, when falling prices triggered systems programmed to sell shares that passed a particular price threshold, which in turn caused prices to fall further, which pushed share prices past yet lower thresholds, which triggered off yet more selling. This time, however, the computing equivalent of fire breaks should have ensured that such automatic trading systems were under control.

One possible cause, however, that the markets did not contemplate – perhaps could not contemplate – was that the 1989 crash was evidence that the world stock markets were, in effect, beyond human control. It has been increasingly hard to relate the movement of the markets to real economic conditions. Japan's Nikkei index, for example, suffered terrible falls in following years, despite the country's robust economic performance. Similarly, the British stock market reached new highs while a persistently weak economy went through one of its periodic slumps.

Though stock markets have always behaved unpredictably, the new electronic, computerized, globally networked markets are far more volatile. There seems to be a diminishing relationship between the value of a company's shares and its actual perform-ance. The 'casino economy', as leftist critics labelled it, was becoming a computer game that was under the control of no one player. Like the patterns in the computer Game of Life, unexpected patterns such as sudden rises and falls in the markets seemed to be the result of the interaction of capital, 'emergent' properties of a complex system rather than the whim of individuals acting according to their best interests.

It was 1930, the year Gödel's famous paper undermining the certainties of mathematics was published, that Britain was forced off the gold standard by increasing instability of the pound, then a major world currency. This total breakdown of the system had only happened twice before since its adoption in the early eighteenth century: during the Napoleonic and the First World Wars.[10] This time it was gone for good, killed off by the world depression that followed the first Wall Street crash of October 1929.

The gold standard was a stabilizing measure designed to provide a fixed measure of money's value by relating it to a fixed quantity of gold. It meant that the purely symbolic nature of paper money was anchored in something that was regarded as having inherent material value. However, for economic reasons, this was a system that the British government could not sustain while the economic order collapsed around it. One might have thought this was the time when something as fixed and certain as gold was worth its weight as a form of security. But, as it turned out, gold itself had no intrinsic value; it, too, could not escape the vagaries of supply and demand on the international marketplace.

The abandonment of the gold standard has turned money into a purely abstract quantity, a symbol. A deficit is no different to a surplus, except that the mathematical sign for one is a minus, for the other a plus. Exchange has become an arithmetical correlation: if my balance goes up by so much, yours must come down by the same amount. Currencies 'float', their level being set by their relationship to all other currencies. There exists no material means of distinguishing 'my' money from anyone else's, no need for physical possession, no stash of gold or even cash in my bank's vaults that increases or diminishes according to how much I earn or spend. In the global village, this process of abstraction has reached its purest expression. Junk bonds, paper billionaires, credit ratings – money is just a parameter in a process running on a global computer, part of the reason that for many it is now meaningless, even worthless, a form of pure abstract symbolism that bears no relation to productive effort or material rewards. It exists in another realm, the same realm as the media virus – in cyberspace.

Perhaps cyberspace, then, is – literally – where the money is. Perhaps it is also the place where events increasingly happen, where our lives and fates are increasingly determined; a place that has a very direct impact on our material circumstances – a blip in the money markets can raise bank lending rates, a blip in a multinational's productivity can close factories and throw economies into depression, a blip in the TV ratings can wipe out an entire genre of programming, a blip in an early warning system can release a missile.

The power of this realm comes from its connectedness. It is a continuum, not a series of discrete systems that act independently of each other. Blips are not isolated events. Accelerated and amplified by electronics, they are the result of the subtle interactions of the millions of bits of data that flow through the world's networks. Furthermore, anyone connected to these networks, be they a humble telephone subscriber or TV viewer, participates in such events. No one can avoid becoming active citizens of cyberspace.

To people who have been involved in computing, and in particular personal computing, in the 1980s, such citizenry is no mere matter of theoretical speculation, it is very real. In 1982, I visited California for the weekly computer trade newspaper *Datalink*. The magazine's editor, Guy Kewney, and I decided to use the trip as an opportunity to try out a new way of communicating text that would avoid the delay and inconvenience of reading the articles over the phone (this was at a time when faxes were still relatively rare, even in California).

I typed up my articles on an Osborne-1 portable microcomputer. About the size of my tightly-packed suitcase, and almost as heavy, it would have probably been easier to lug around an IBM golfball typewriter, but I needed to have my text in 'machine-readable', word-processed form, not typed on paper, because I was going to send it back to Britain in that form. To do this, I used a modem – a device that turns the electronic signals generated by a computer into the sort of signals carried by a telephone line – and dialled up an 'electronic mail' system in the UK called 'Gold'. Once connected, I sent the file that contained the text of my article down the line to Gold, which filed it away in its own computer for retrieval by colleagues at the office.

The result was instant, international communication, and, in the following few years, it became a standard means for reporters to file their copy, particularly so as the weight of portable computers decreased. It was just one of a growing number of ways that computers and telecommunications combined to change working practices. Academics used the ever-spreading networks to communicate with their peers by electronic mail. Businesses began to dial up databases of stock market prices, press cuttings or company accounts. Indeed, by the late 1980s, a whole new multi-billion dollar 'value added services' communications market had emerged. Businesses like AT&T were no longer telephone companies, they were 'telecommunications' companies, their networks were no longer just for voice, they were for any kind of data.

Computer users were plugged straight into this growth in telecommunications. Many linked up to the ever-increasing number of electronic services at their disposal: bulletin boards, online databases, home banking, teleshopping. The experience of using such services powerfully reinforced the collective imagination of computer users that there was another 'world', a world where much of their social intercourse might take place, where much of their information would come from. They also knew that it was linked into the networks of international finance, commerce and government – as demonstrated by the growing number of stories about hackers gaining unauthorized access to company and defence systems using the same basic equipment. And it was the world infected by the computer virus, spread by the exchange of pirated and 'homebrew' software deposited on bulletin boards and conferencing systems.

It was the excitement of being part of this world that stoked up the computing community's interest in virtual reality. Could this be where the denizens of the global village truly belonged? Could this be a *new* reality?

Notes

1 Interview with the author, *Late Show*, BBC 2, 26 September 1990.
2 William Gibson, *Neuromancer*, London: Grafton, 1986, p. 67.
3 David Gans and R. U. Sirius, 'Civilizing the electronic frontier', *Mondo 2000*, 3, Winter 1991, p. 49.
4 Marshall McLuhan, *Understanding Media: the extensions of man*, London: Routledge, 1964, p. 5.
5 Mark Poster, *The Mode of Information: poststructuralism and social context*, Cambridge: Polity Press, 1990, p. 2.
6 Wes Thomas, 'How I created a media virus', *Mondo 2000*, 2, Summer 1990, p. 137.
7 Richard Dawkins, *The Selfish Gene*, 2nd edition, Oxford: Oxford University Press, 1989.
8 Richard Dawkins, *The Extended Phenotype*, Oxford: Oxford University Press, 1982, p. 200.
9 Stephen Kindel and Amy Barrett, 'The crash that wasn't . . . or was it?, *Financial World*, 14 November 1989, 158 (23), p. 26.
10 E. J. Hobsbawm, *Industry and Empire*, London: Penguin, 1969, p. 236.

7
INTERFACE

'Long ago a Western poet, the noble Xi-mo-ni-de, was gathered with his relatives and friends for a drinking party at the palace, among a dense crowd of guests. When he left the crowd for a moment to step outside, the great hall came tumbling down in a sudden mighty wind. All the other revellers were crushed to death, their bodies were mangled and torn apart, not even their own families could recognize them. Xi-mo-ni-de, however, could remember the exact order in which his relatives and friends had been sitting and as he recalled them one by one their bodies could be identified.'

    Jonathan D. Spence, The Memory Palace of Matteo Ricci

This tragic tale is the version of a story found in Cicero's *De Oratore* as retold by the Italian missionary Matteo Ricci to the Chinese in the late sixteenth century. 'Xi-mo-ni-de', Ricci's Chinese rendition of a name translated into English as 'Simonides', was a lyric poet born on the Greek island of Ceos in 556 BC. His mysterious ability to recall the seating plan of the entire party is the first recorded example of the use of a memorizing technique that was to form the basis of the art of rhetoric and, some might argue, Western literature. We now think of rhetoric as something deployed by politicians to fool those who listen to them. But, before the development of printing and the spread of literacy, it was the main art of communication, the means by which story tellers could learn how to tell their stories in a way that was convincing and effective.

In pre-literate, oral cultures, there was no means available to most story tellers to record the tales they told other than by committing them to memory. Many stories have a simplicity that makes their memorization easy – fairy tales are an obvious example. However, some stories are simply too long or too complicated to be remembered without some sort of aide memoire. Orators had to be able to recall stories like Beowulf, which emerged in the tenth century as a 3,200-line poem with a plot that would daunt a modern-day thriller writer.

The technique that enabled Simonides to perform his memory feat was the one he used to tell his stories, whereby he would think of a sequence of ideas as a space, and populate the space with objects that would jog his memory. By mentally walking through this imaginary space, he was then able to recall the sequence of ideas. Obviously, the habit helped him perform the same trick when physically walking round the remains of the unfortunate guests. Rhetoricians referred to this imaginary space as a 'memory palace', and it formed the basis of mnemonics right up to the time that Matteo Ricci attempted to convert China to Catholicism.

According to the historian Jonathan Spence, who wrote a definitive account of Ricci's mission, Ricci identified three types of imaginary locations that could make up a memory palace: they could be based on reality, on buildings or objects that the orator had actually encountered; they could be fictional; or they could be half real and half fictional – for example, a real building could be imagined as containing a staircase or doorway that lead to imaginary floors or rooms. Spaces created out of such combinations of the real and imaginary could, Ricci claimed, become 'as if real'.[1]

Ricci's memory palace took the form of a series of images based on Chinese ideograms and pictures representing scenes from the Christian story that would be arranged in a series of rooms as paintings in a gallery. However, memory palaces could take many forms. Medieval churches were often designed to be simple mnemonic spaces that reminded worshippers of the Christian story, perhaps with a sequence of frescoes depicting Christ's Passion along the walls of the nave. In combination with strategically placed objects – the font near the entrance, the memorials of past bishops surrounding the altar – an entire cathedral became a metaphorical representation of the pilgrim's progress from birth through to death and salvation.

'The idea that memory systems were used to "remember

Heaven and Hell"', wrote Jonathan Spence, 'can explain much of the iconography of Giotto's paintings or the structure and detail of Dante's Inferno, and was common-place in scores of books in the sixteenth century.'[2] It is arguably built into the foundation of literature as a whole, the idea of fiction as a metaphor for reality, a way of exploring it and making sense of it. The memory palace was, at least from the point of view of the preacher, a way of modelling, as it might now be called, the reality of religious experience. Metaphor, the keystone of literature, the 'lie' that can give us access to the 'truth', perhaps has its origins in a mnemonic device.

In 1976, Nicholas Negroponte of the influential Architectural Machine Group began to work on what he and the co-researcher Richard Bolt called 'spatial data management'. Negroponte's idea was to turn the computer into a memory palace – or, more precisely (and prosaically), a memory office. The computer screen would become a metaphorical desktop covered in objects representing various functions or tools: diary, telephone, 'in' tray for important memos and so on. The concept of spatial data management was developed by the Architecture Machine Group in an attempt to discover ways of placing the user within this 'office'. The group created what it dubbed a 'media room' described by Stewart Brand, in his book on the MIT Media Lab, as a 'room-sized personal computer where the whole body [is] the cursor-director and your voice the keyboard'.[3] The user sat on a chair in the middle of the room facing a wall upon which was projected a computer-generated image – the 'space' representing whatever data needed to be manipulated. The user, whose arm movements were tracked by sensors connected to the computer, could select objects displayed on the wall by pointing at them. While pointing at a target object, he or she could say 'put that . . .', then point to elsewhere on the wall and intone '*there*' to move the object from its old position to a new one.

Though perhaps the most influential in terms of the subsequent development of virtual reality, Negroponte's attempt to create a computer-generated environment was by no means the first. In April, 1969, an exhibition opened at the Memorial Union Gallery at the University of Wisconsin which required visitors to walk into a dark room with tubes of fluorescent light lining the walls. As they walked, Moog music would rotate around the room, and waves of coloured light would pass down the tubes, producing a response described by one of the exhibit's creators, Myron Krueger, in terms that very much capture the spirit the times: 'People had amazing reactions to the environment. Communities would form among strangers. Games, clapping, and chatting would arise spontaneously. The room seemed to have moods, sometimes being deathly silent, sometimes raucous and boisterous. Individuals would invent roles for themselves. One woman stood by the entrance and kissed each man who came in, while he was still disorientated by the darkness.'[4]

Krueger described the installation, called 'Glowflow', as a 'kinetic environmental sculpture'. More significantly, it was a prototype of a technology that Krueger was later to call 'responsive environments'. In Glowflow, pressure pads in the floor would activate sequences of sounds and lights when they were pressed, providing a crude sort of response from the artwork. Krueger refined this idea with his second installation, Metaplay. Visitors would enter a room and see before them a wall with their live video image projected upon it. Superimposed on this image would be a line drawing produced by an artist who, watching the same video image as the visitors, could change the drawing in response to their actions.

What interested Krueger about the experiment was the visitors' response to their own image and to its manipulation by another. It was this that he wanted to explore further with what he regarded as the first truly responsive, artificial environment, Videoplace. The idea behind Videoplace was to create a space that reproduced the experience of meeting in a physical space, an artificial world where people could see, hear and touch each other, see and manipulate the same objects.

Krueger originally sought funds to develop Videoplace as a global meeting place that would commemorate the US Bicentennial in 1974. The reaction he got from the scientific funding bodies – the US National Science Foundation, DARPA (which two years later would be funding Negroponte's own version of the responsive environment) and NASA (which said he could use one of their satellites, as long as he could get it launched) – was not encouraging. Orthodox science-funding bodies seemed to be unable to take an artist with an engineering bent seriously. Krueger has since expressed some anger at this, which is understandable given that it took another decade for the rest of the computer world to catch on to the idea, though even with the benefit of hindsight, it is difficult to see how a funding agency could justify backing a project on the basis of a track record which comprised a few slightly flaky art installations.

In any case, Krueger eventually found a home for his ideas at the Computer Science Department of Connecticut University. There, he built an installation based on Metaplay, with users facing a screen upon which was projected a live image of their silhouette. The difference was that the image was processed by a computer, which 'recognized' the silhouette shape, and therefore could generate its own graphics that would interact with it. So, users could, say, 'touch' a computer graphics object in the image they saw projected on the screen, and 'move' it. The projected picture became a sort of two-dimensional image of an environment in which the user – or rather the user's shadow – was immersed and with which the user could interact. By linking installations together, several environments could be merged into one, so the user could see the image of a user at another installation in their own environment and thereby 'meet' that user.

Complex though it may sound, the beauty of the idea was its relative technological simplicity. Research into Sutherland's concept of the ultimate display relied on creating a whole suite of new technologies: the head-mounted display, the Dataglove, position and movement sensors, actuators. It also required immense computing power, since it depended on the generation of three-dimensional graphics in real time, in response to the user's movements. Krueger's responsive environments could be created using little more than a standard video camera and a home computer. 'The idea that people are going to put on gloves and scuba gear to go to work in the morning at least requires some scepticism,' he later told a computer graphics conference,[5] and he had a point.

Whatever the best way to implement it, the importance of the idea of using the computer to create a metaphorical environment or memory palace has proved to be one of the most influential in computer design since the emergence of personal computing in the early 1980s.

Before the personal computer era, computers were specifically designed for use by trained operators and specialists. The machines and their minders sat in air-conditioned computer departments, insulated physically as well as organizationally from the institutions they were there to serve. The 'user' was expected to know nothing about the system, only what it could do – it was up to the professionals to work out how to get the computer to do it.

However, as the cost of the hardware started to plummet with the development of microchips, so did the expertise of the users. It made no sense to employ a specialist to operate a computer that cost only a few thousand pounds and that was to be used by only one person. It was unreasonable to expect everyone who wanted to benefit from the much-heralded computer revolution either to be a data-processing expert or be prepared to employ one. It was especially unreasonable given that most people were scared of the technology, more inclined to associate it with the police state and dehumanizing automation than personal empowerment.

That was why IBM launched its Personal Computer with an advertising campaign that deployed the most human, most familiar imagery it could think of. It wasn't a work tool on offer, it was a friend, in the form of Charlie Chaplin (or rather, an actor impersonating him) proffering a rose. 'Friend' became the key word of personal computer marketing, as in that awkward term 'user friendly'.

But, to begin with, PCs were anything but friends. Time and again, it was shown that users were employing a tiny proportion of the available facilities. They used word processors as they would type-writers, databases as they would manual filing systems, spreadsheets as they would calculators. And software designers designed their personal computer programs accordingly – word processors behaved like typewriters, databases like card-indexing systems – which in turn diminished the reason for buying the computer in the first place.

The problem lay in what came to be called the 'human–machine interface', the point of contact between the user and the machine. In old-fashioned mainframe computers, the interface was impossibly complicated, involving innumerable stages and people deployed at a pace set by the technology rather than the user – in 'batch' mode, as it was called, after the need to do jobs in batches large enough to smooth out peaks and troughs in the demand for computing time. With the development of minicomputers in the 1970s, 'real time', 'interactive' systems emerged that would perform operations at a pace set by the user. This became possible not simply because the hardware became more powerful; it was the result of the development of sophisticated programs called 'operating systems' that automated a lot of the administrative tasks of running the computer itself, previously performed by engineers. Such operating systems allowed the user to 'interact' directly with the system, type in instructions via their own keyboard, rather than via pre-printed forms that had to be processed by the computing department, and see the results displayed on their own screen or printed out on their own printer the moment they were available.

These more advanced operating systems used what came to be called a 'command line interpreter' (CLI) as the user interface. Communication with the system was a written correspondence of often obscure orders with a very pedantic servant. But at least the servant was, organizationally speaking, more or less under the user's direct control. Also, the servant could, theoretically speaking, turn the sorts of commands that a non-technical user might issue into the appropriate machine instructions.

It was in the world of the minicomputer CLI that most of the 'pioneers' of personal computing cut their teeth, so it was unsurprising that this was to inspire the design of the first PC operating systems. 'CP/M' (its initials originally stood for 'control printer/monitor', but was later revised to stand for 'control program for microcomputers'[6]), the first mainstream program, and the CP/M-derived 'MS-DOS', the operating system adopted by IBM when it launched its PC, were both classic CLI interfaces.

CP/M and MS-DOS were primarily adopted by microcomputer manufacturers because they were designed to be used with the then most popular microprocessor, the Intel 8080, and could be easily adapted to different hardware configurations. They were not adopted because they were particularly easy to use. CP/M was notoriously difficult. The command to copy a file from one device to another (say from one disk to another) was PIP. If anything went wrong, all the user would get was the less than helpful message: BDOS ERROR ON A. MS-DOS was little better. PIP became COPY, which helped, but still the user was expected to engage in this strange, stilted written communication with some alien, disembodied 'thing' that was anything but a Chaplin.

* * *

In 1970, the Senate passed what came to be known as the Mansfield Amendment, after the Democratic Senator who moved it, Michael Mansfield. Mansfield, worried about the increasing influence of the Pentagon over academic research in non-military fields, proposed an amendment to the Defense Appropriations Bill requiring that all projects of the Advanced Research Projects Agency (ARPA, the agency that set up ARPAnet) be defence or 'mission' orientated.[7] This stopped at a stroke a key source of funding for a wide variety of civil projects inspired by computing's first wave 'visionaries' and 'mavericks', men like the one-time radar technician Doug Engelbart and the psychologist J. C. R. Licklider.

Licklider created and Engelbart aimed to realize what came to be called the 'ARPA dream' of turning the computer into a machine that could, in Engelbart's language, 'augment' the human intellect just as other machines had augmented the human body. Licklider wanted to discover a way to bring humans and computers together into a new sort of symbiosis. 'There are many man–machine systems. At present, however, there are no man–computer symbioses,' he wrote in 1960. 'The hope is that, in not too many years, human brains and computing machines will be coupled together very tightly, and that the resulting partnership will think as no human being has every thought.[8]

In 1962, Licklider was given a post at ARPA to search for ways of achieving this symbiosis, and immediately set about distributing the vast sums available to him. He fuelled a great national discovery of interactive computing, which reinforced the conviction that, outside the data-processing departments of telephone and electricity companies, there was no room for the constipated pace of batch-drive mainframes.

Among the projects he backed was Engelbart's Augmentation Research Center, ARC, at Stanford University, which was set up to explore new forms of computer interaction, developing, among other innovations, the mouse that has since become one of the standard devices of personal computing. He also supported the

work of a student called Alan Kay, who was working on Ivan
Sutherland's project to find ways of using pictures to depict the
ever-changing information created by interactive systems on
computer screens.

In 1970, Xerox, the company set up to exploit the photocopying
technology IBM had abandoned in the 1950s, acquired a new
Harvard-trained chief, Peter McColough. Xerox was concerned
about the threat the computer represented to its core business – it
was understandably nervous at all the talk about 'paperless'
offices – and wanted to cover its options by developing itself into
an 'information' company. So it set up its interdisciplinary Palo
Alto Research Center (PARC) near the Stanford University
campus with the aim of producing a response to IBM's dominance
of the information industry.

The timing was fortuitous – it was the same year as the ARPA
researchers were having their dream rudely interrupted by the
realities of military funding and the Mansfield Amendment. So
Xerox sensibly hired a former ARPA projects manager, Bob Taylor,
to help recruit staff to the center. Alan Kay was made chief scientist,
and he, along with former Utah and ARC scientists such as Larry
Tesler, set about working with hardware engineers to create a truly
interactive computer called the Alto. This was to be the computer
that would break the mould by abandoning the CLI interface in
favour of a revolutionary new one based on the ARPA dream.

Tesler later explained in an interview for the BBC's history of
computing, *The Dream Machine*, how this new interface would
work: 'What we realized was that we could create what some
people called a user illusion, something that appears to be a world
on a screen. One way to thinking about it is you play a video
game . . . there's an illusion of spaceships or roads and cars . . .
and the user who gets engrossed in the game . . . starts operating
as if they're really working in the real world . . . when in fact
they're only working in this imaginary, simulated world created
by the sequence of steps in the computer program.'[9]

The approach the PARC researchers adopted was to show the
computer's resources graphically. They did this by depicting
what it could do by means of metaphors. A famous example of
this was a simulated painter's palette, created by Kay in 1972,
which enabled the user to 'paint' pictures on the computer screen.

This was just a start. They did not want simply to create metaphorical tools, they wanted to create a whole metaphorical world, one in which the user would explore to discover what the computer could do. The command line would be replaced by a highly abstract picture or map of a metaphorical office, with objects within the office such as storage devices (the 'filing cabinets'), files and tools represented by small pictures called 'icons' depicting some aspect of their function. The result was a more businesslike version of the environment created by computer games like Space Invaders or Missile Command. In such an environment one object becomes a doppelgänger 'you', just as the piece you have control over in a board or video game becomes 'you'. In the case of Xerox's environment, your doppelgänger is a pointer which roams through the virtual office in accordance with movements that you make using the 'mouse' that the PARC researchers had brought along with them from Engelbart's ARC. By using this pointer to point at 'objects' in the environment represented by the icons, they can be manipulated.

While this research work was going on, the first microcomputers were beginning to find their way onto the market, and in 1979 Xerox decided to invest in Apple. The decision was, at one level, eminently sensible: Apple went on to become one of the most spectacular success stories in post-war business history. But there was a price. In order to get hold of the much sought-after Apple shares, Xerox had to offer a deal: in exchange for 100,000 of them, it would give Apple access to PARC's research.[10] The deal was to prove expensive.

Due to the departure of Peter McColough and the new threat of Japanese competition, Xerox's managment decided not to bother turning the Alto into a marketable product. So, it was just lying there, neglected, when Apple's representative, the company's co-founder Steve Jobs, arrived to look round the research center. At PARC, Jobs found, as Frank Rose put it in his book about Apple, *West of Eden*, the 'Hanging gardens of the information age: a terraced structure built into a golden hillside high above the bay, filled with electronic marvels of tomorrow'. He also found the Alto. It was, perhaps, no surprise that a member of the first television generation should have found this picture machine so exciting. More surprising is the fact that Jobs was allowed to develop the technology, worth millions of dollars, into a market-able product for free.

The result was the Apple Lisa. I first encountered one of these in the UK in early 1983. It was a machine shipped in direct from America. The power supply was the US standard 110 volts, so a transformer had to be used to adapt it to 240 volts; there was no documentation or software apart from a box containing a couple of illegibly labelled floppy disks; the casing was damaged because it had been delivered without packaging. But I did not mind. It was the only one in the country.

The arrival of this machine had been anticipated with more excitement than even the IBM Personal Computer, which had been introduced only two years earlier. Even its name had aroused excitement – Lisa was, according to a report in *Time* magazine, the name of a girl cited in a paternity suit against Jobs, while Apple unconvincingly claimed that it was an acronym for Locally Integrated Software Architecture. Whatever the truth (it seems likely to have been the name of a daughter of one of the design engineers), it was proclaimed a revolutionary machine. The Apple II took 20 hours to learn, said Apple; 'Lisa takes 20 minutes'.[11]

So, notwithstanding the absence of any sort of manual, I set about seeing what I could learn in the allotted time. Soon after switching on, I saw what was then the revelatory sight of a 'bit-mapped' display, the full graphic glory of Lisa's 'desktop'. I quickly learnt how to use the mouse to perform simple operations, such as 'picking up' an icon depicting a folder marked 'blank paper', 'dragging' it across the screen and 'dropping' it into another little icon entitled 'rubbish bin'. There were pictures of a clock, a calculator, both fully operational, plus icons depicting filing cabinets and stationery cupboards.

But there were snags. The floppy disks the Lisa used were quite new: they had far more than the usual number of notches in their sleeves, which made it impossible to tell just by looking at them which way they slotted into the drive. So I guessed, and guessed wrong. In any other system, this would have presented no great problem, since the disk could be removed and reinserted the right way. Unfortunately, the Lisa had no eject button for removing the disk; indeed, I quickly discovered that it did not even have an off button. Both were operated by the system itself, and it was by no means clear how the machine could be instructed to release its disk or switch off.

The experience – a very unpleasant one, given the scarcity of the machine at the time – was instructive. Certainly, the system had been easy to use within clearly defined limits, but if it was a 'world on a screen', in Tesler's phrase, it was one that lost its consistency pretty quickly. It also put paid to the idea that the user was somehow more in control. It was the machine that was deciding, though no doubt for my own good, what should happen when. There was something quite sinister about not even being 'allowed' to switch the thing off, forcing me to contemplate either pulling the plug, an option which the machine left you in no doubt would do more harm than good, or just leaving well alone. It was hardly an experience of the individual empowerment that the personal computer revolution was supposed to be offering.

Nevertheless, the Lisa was a breakthrough. It was the prototype of both Apple's hugely successful Macintosh, and inspired Microsoft Windows, a version of the interface designed to run on IBM PC-type machines. Indeed, many of the minicomputer systems used by specialists now employ these 'graphical user interfaces', GUIs. Xerox, too, has continued to develop the idea, introducing a simulated third dimension to the previously 'flat' desktop environment to create 'rooms' connected by 'doors', allowing the user to move from room to room, carrying any tools they need in their 'pockets'. Though the user only sees these rooms projected into two dimensions on a flat screen, the dopplegänger pointer is movable in three.

This, for the virtual realists, is what cyberspace is about, because ultimately their technology offers the possibility of a union between users and their virtual dopplegängers – a true state of human–machine 'symbiosis'. Such environments will become accessible 'for real', making the user interface disappear and immersing us in the universe of information. And what a universe! It will be, the virtual realists promise, a realm far greater and richer than the physical one, a realm that we have so far only dimly perceived through our imaginations, a realm that artists have struggled to reveal using mute, passive media such as paint and paper.

Notes

1 Jonathan D. Spence, *The Memory Palace of Matteo Ricci*, London: Faber, 1985, p. 2.
2 Spence, 1985, p. 13.
3 Stewart Brand, *The Media Lab*, London: Penguin, 1989, p. 139.
4 Myron W. Krueger, *Artificial Reality*, Reading, Mass.: Addison-Wesley, 1983, p. 15.
5 'Hip, Hype, Hope' Panel, SIGGRAPH 90, Dallas, Texas, 10 August 1990.
6 Christopher Bidmead and Benjamin Woolley, *The Micro Enquirer*, London: Century, 1984, p. 52.
7 Frank Rose, *West of Eden*, London: Hutchinson, 1989, p. 45.
8 J. C. R. Licklider, 'Man-computer symbiosis', *IRE Transactions on Human Factors in Electronics*, vol. HFE-1, March 1960, pp. 4–11.
9 Quoted in Jon Palfreman and Doron Swade, *The Dream Machine: exploring the computer age*, London: BBC, 1991.
10 Rose, 1989, p. 47.
11 Trevor Huggins, 'Desktop display', *Computer Answers*, March–April, 1983, p. 11.

8
HYPERTEXT

'Its shelves register all the possible combinations of the twenty-odd orthographical symbols (a number which, though extremely vast, is not infinite): in other words, all that it is given to express, in all languages. Everything: the minutely detailed history of the future, the archangels' autobiographies, the faithful catalogue of the Library, thousands and thousands of false catalogues, the demonstration of the fallacy of those catalogues, the demonstration of the fallacy of the true catalogue, the Gnostic gospel of Basillides, the commentary on that gospel, the commentary on the commentary on that gospel, the true story of your death.'

And this book.

All these are to be found within the Library of Babel, a library of incomprehensible immensity imagined in a short story by Jorge Luis Borges.[1] The library comprises 'an indefinite and perhaps infinite number of hexagonal galleries', each containing 20 shelves, each shelf containing 35 books, each book containing 410 pages, each page containing 40 lines, each line containing 80 letters. This is not a library to discover a long-lost literary oddity or neglected work of genius. The books have no title or author; they are simply arbitrary collections of symbols, each one a combination of all the possible combinations of the letters of the alphabet. That is why the library is so vast and why it contains every possible text, including this one.

There may be no real Library of Babel, but its books do, in some sense, exist as part of the information realm. They exist in the same sense that the number at the billion billionth decimal place in the mathematical quantity pi exists. The true story of your death exists in that it is a particular configuration of alphabetic characters. The problem is that it is one of many. It cannot be found by sifting through the entire Library of Babel volume by volume, not in all the history of time. But it is, perhaps, possible to discover texts that no individual author has so far found, variations on existing texts, alternative endings to familiar scenarios, elucidations, glosses, diversions, a new text that can be discovered by its proximity to an existing one. That, at least, is the hope of the champions of a new sort of literary object: the 'hypertext'.

Now what do you want to do? The choice is yours. You could, for example, read on. Or you could tear the pages of the book out, even cut them up into tiny pieces, and reassemble them in a new order. Or you could write the rest of the book for yourself. Or, in a few years time, when the book is available in electronic form, you could combine it with other electronic texts and thereby create a book of your own, perhaps for publication, perhaps for your own personal benefit.

I have been exceptionally generous in giving you these choices. Books, with the obvious exception of reference sources and suchlike, are written to be read in the order and fashion set out by the author. This is a principle enshrined in the Bible, the book whose title simply means book, containing the Word of the greatest author of all, God. Books are not, according to the classical view of them, the result of collaboration or negotiation; they are given by their authors so that they may be taken by their readers. They are also the model of all other cultural artefacts. paintings, plays, films, TV, buildings. In each case, the viewer, audience, inhabitant, whatever, is part of a normally non-negotiable pact with the painter, director, producer, architect, whoever, a pact which involves producing on the one part and consuming on the other.

With the development of interactive computing, this classic author/reader distinction is looking less and less valid. Nicholas Negroponte's prediction that 'Prime time will become my time' expresses the hope that technology can destroy the tyranny of Biblical authority.

This hope is a product of the shift of the computer interface away from what might be called the mechanical self model to the cyberspace model, from the command line interpreter's keyboard conversation between human and machine towards the mouse-directed exploration of artificial environments. Much of the computing world has, at least at the time of writing this fixed, non-interactive text, failed to recognize this shift. The power of the old-model artificial intelligence lobby still prevails. Finding a way of using the computer to reproduce intelligence remains the preferred research strategy of the defence industry – witness the emphasis it placed on the 'smartness' of the weapons deployed in the Gulf War, and the implication that AI had somehow created a new sort of limited, 'surgical' battle and a new sort of discriminatory, almost moral class of weapon. In universities, too, AI remains the computing department's Grail, perhaps because it represents the chance to achieve an almost alchemical transition of dead machinery into a thinking being.

Nevertheless, a whole host of computing developments, such as scientific visualization, the study of chaos, fractals and 'hypermedia' have opened up a new agenda for computing's research community, one that sees interactivity not in terms of a conversation between a natural and an artificial intelligence, but as an exploration of some form of cyberspace. 'I believe that conversation is the wrong model for dealing with a computer,' wrote John Walker, the enigmatic founder of the Californian software company Autodesk. 'Since inception we've seen computers as possessing attributes of human intelligence ("electronic brains"), and this has led us to impute to them characteristics they don't have, then expend large amounts of effort trying to program them to behave as we imagine they should. When you're interacting with a computer, you are not conversing with another person. You are exploring another world.'[2]

'The method is simple,' wrote William Burroughs, regarded by many virtual realists as the proto-cyberpunk novelist. 'Here is one way to do it. Take a page. Like this page. Now cut down the middle and across the middle. You have four sections: 1   2   3   4 . . . one two three four. Now rearrange the sections placing section four with section one and section two with section three. And you have a new page.'[3] This is the cut-up method of writing, which novelists like Burroughs developed to destroy the control of narrative. Why, he and those like him wanted to know, should z follow y, day follow night, two follow one, hangover follow reckless indulgence? The apparently 'natural' sequences are, according to Burroughs, just one set of progressions among many, your set, perhaps, but not necessarily mine.

The idea of the cut-up or collage method had its origins in dadaism. In the 1920s, Tristan Tzara, who inspired Burroughs's experiments into narrative disintegration, developed a technique for writing poetry by pulling a poem out of a hat – literally. He would jumble up some words written on slips of paper, and take them out of a hat one by one to see what poem resulted.

Burroughs, in the language typical of the manifestos that pre-war art movements had a taste for publishing, declared: 'Cut-ups are for everyone. Anybody can make cut-ups.' It was this language of liberation that was to be picked up by the computer revolutionaries. The computer could do your cut-ups for you, and stick them back together so seamlessly that you would not even know that they had been made. That was the promise of the 'interactive' novel, the artefact that the computer would enable users to create for themselves. Literature is for everyone. Anyone can create literature.

The underlying principle of the interactive novel is not that the computer becomes the user's pet author, writing books that suit its owner's particular tastes, but that the imaginary world of a novel, the world through which the novel treads just one path, the path of its narrative, can be turned into a virtual world. The computer, then, provides the means of navigating this world. The user flies through it as he or she might fly through an artificial landscape in a flight simulator, plotting a path that is, in effect, the narrative of their own, personalized novel.

The technological challenge of interactive fiction was to find a way of turning imaginary worlds lodged in the writer's head into virtual worlds lodged in the computer's memory. When the idea was first being developed in the mid-1980s, there was a precedent that seemed to promise a way forward: the 'Adventure' game.

Adventure was developed in the 1960s at SAIL, Stanford University's Artificial Intelligence Laboratory. As Steven Levy points out in his book *Hackers*, being in California, SAIL was very different in character to the famous AI Lab located in MIT's stark, strip-lit Tech Square building: 'Instead of the battle-strewn imagery of shoot-'em'up space science fiction that pervaded Tech Square, the Stanford imagery was the gentle lore of elves, hobbits and wizards described in J. R. R. Tolkien's Middle Earth trilogy.

Rooms in the AI lab were named after Middle Earth locations, and the SAIL printer was rigged so it could handle three different Elven type fonts.'[4] A computerized version of 'role playing' games like Dungeons and Dragons, Adventure comprises a series of descriptions of fictional locations inspired by Tolkien and the surrounding Californian mountains. The user navigates these more or less 'geographically', by typing in instructions that specify which direction he or she would like to go in (which might take the form of 'go north' or 'step back') or the actions he or she would like to take (such as 'pick up axe').

Though 'just' a game, playing it is a bit like reading a novel. Many of the pleasures of playing are literary ones, coming from the rich descriptions (too rich, perhaps, for some tastes) of the locations encountered, and the sense of discovering a coherent fictional world. 'You are in a splendid chamber thirty feet high', reads the description of one location. 'The walls are frozen rivers of orange stone.' 'You are at one end of a vast hall stretching forward out of sight to the west,' reads another, 'the hall is filled with wisps of white mist swaying to and fro almost as if alive.' And, the most famous of all, reached after spending days lost in mazes and being attacked by dwarfs and trolls: 'Far below you is an active volcano, from which great gouts of molten lava come surging out, cascading back down into the depths. The glowing rock fills the farthest reaches of the cavern with a blood-red glare.' It was this literary aspect of the game that inspired researchers to look for a way of creating interactive novels, 'hypertexts' that could be read in an order that suited the reader, rather than along the linear passage of the author's narrative.

Michael Joyce, coordinator of the Center for Narrative and Technology at Jackson Community College in Michigan and J. David Bolter, computer academic and author of the book *Turing's Man*, set about trying to achieve this with a program they called Storyspace. Storyspace was an 'authoring' tool, providing a means of drawing together texts, structuring them and navigating between them. With a conventional word processor, a document is a linear block of text that has a beginning, a middle and an end. With a hypertext system like Storyspace, it is a series of units – in Storyspace, called 'spaces' and 'places' – which are structured into some sort of network. So, for example, rather than one sentence leading on to the next, it may, by selecting from a series of navigational tools, lead laterally on to several other related sentences, depending on the reader's response to it.

Joyce used Storyspace to build an interactive novel, as well as adapt Borges's *The Garden of the Forked Paths*, a story which itself (like so much of Borges's work) was an attempt to explore the dimensions and structure of imaginative space. Reading Joyce's interactive novel was an intriguing, and often disorientating, experience. On the plus side, he demonstrated that it was possible to create texts which did not force their readers down one particular route, like tourists being chaperoned round a country house. Readers could flout 'No Entry' and 'This Way' signs, plotting a route that suits them. On the minus side, they stood a very serious risk of getting totally lost, perhaps partly because the 'landscape' created by the text was an unfamiliar one, partly because the narrative is the means by which readers orientate themselves.

Joyce and Bolter recognized such difficulties, and sought to overcome them by placing limits on narrative freedom. 'Storyspace', wrote Joyce, 'might . . . be used to create a novel as supple and multiple as oral narratives.'[5] In other words, like a story told by word of mouth rather than fixed in print, such a novel could be retold as a storyteller might retell a favourite tale, in many different versions, elaborating on different parts to reflect the mood of the audience. At the same time, it would be a novel with the 'referential and coherent richness' of the sort written by James Joyce, one that created such a plethora of meanings that it was up to readers to decide for themselves how to make sense of it.

The main problem with the concept of the interactive novel is its assumption that narrative is in some sense independent of the imaginary realm it navigates. You could argue that it is the story that creates the imaginary realm, and gives the fiction meaning. A story is what distinguishes random experiences from meaningful ones. In his book on post-modernist fiction, Brian McHale cites John Fowles's famous novel *The French Lieutenant's Woman* as an example of how contemporary fiction plays with the idea of narrative.[6] Fowles gives the book two endings (three, if you count the one three-quarters of the way through, which supplies a conventional Victorian outcome), which in Karel Reisz's and Harold Pinter's film adaptation is ingeniously translated into a film-within-a-film. However, in doing so, he was not inviting his readers to find out their own authentic history of the French Lieutenant's Woman, he was highlighting the spurious meaningfulness of the fictional world he had created, and the indeterminacy of the real one.

* * *

'On bad days I feel I could have saved the world if I had only been more efficient', Ted Nelson told *Dr Dobbs Journal*, the American personal computing journal.[7] One hardly dares imagine what he feels on good days. The reason for his frustration was the inadequacy of books. He felt that in hypertext and hypermedia, terms he is credited with coining, he had the basis of a whole new type of publishing medium, one that would change the way books and other texts – indeed, all sorts of media – are produced and consumed. This new medium would become a text repository, even a vast database of the corpus of English literature, and it would be called Xanadu.

However, since coming up with the idea in 1960, things have not gone entirely to plan. The book has dodged the demise he predicted for it by 1962, and the Xanadu project was not completed in the six months he originally allocated. 'I mistook a clear view for a short distance,' he later admitted.[8] Like a preacher impatiently awaiting the arrival of Judgment Day, he was still predicting that Xanadu, or at least the software needed to build it, was just six months away in 1990. Six months after he said it (I was there; I heard him), it still was not, but by that time most who had heard about it hardly cared: it was the concept that mattered, not its realization.

What Nelson imagined was a not quite the pleasure dome of Coleridge's poem, though that was his inspiration. It was based on an idea first suggested by Vannevar Bush, previously scientific adviser to President Franklin D. Roosevelt. In an article published in *Atlantic Monthly* magazine in 1945 entitled 'As we may think', Bush described a desktop apparatus which he called the 'Memex', comprising a slanting translucent screen 'on which material can be projected for convenient reading', a keyboard, and sets of buttons and levers.

The Memex machine – never actually built, at least in the form described by Bush – would probably have been described at the time Bush came up with it as an automated mechanical reading device. Data was stored on microfilm, then the state-of-the-art in miniaturization, and particular texts called up and projected on the screens by pressing a particular configuration of keys. Thus far he had described a not particularly interesting automated library reading desk. However, he also envisaged the levers and buttons being used to create 'trails' between different texts. These trails would comprise a series of addresses, each one identifying the location of a specific text relating to a particular subject.

Bush's key observation was that Memex users could create their own documents out of the reams of text stored in the machine. This would be done not by laboriously copying the contents of each screen containing relevant text and combining the results to form a new physical document, but simply by keeping a record of the trail, the list of addresses used to access the relevant data. Furthermore, trails could themselves be combined and manipulated to produce different documents on related subjects. The trails, in other words, became the document.

A computer is a perfect Memex machine, being able to handle large amounts of information, display it on a screen, and provide the means of creating trails that link relevant bits of information together. But it could provide further benefits. For example, word processing would enable the text not just to be read, but to be changed or annotated. The computer can search for words and phrases, access distant sources of information over a network, and combine pictures and sound with text.

The Xanadu project aimed at exploiting these advantages. It was to comprise two main components. A company called the Xanadu Operating Company (XOC), owned by Autodesk, was given the task of producing what Nelson called 'a hypermedia server program'. This would provide the mechanism for exploring large computerized databases of information comprising video, music and voice as well as text. This 'server' would also provide the means of creating new documents by establishing links between the content of existing ones, so that each user would be able to establish his or her own view of the database's contents. These links are the key to the Xanadu concept, since it is through links, through the creation of structures within a vast, shapeless mass of information, that Xanadu creates new meanings and interpretations that would be inaccessible using conventional methods of information storage.

The second component of the project – by far the most ambitious – was to create a completely new publishing market. This would comprise a network of databases to which publishers would be invited to contribute the data that would gradually accumulate into a global information repository. Users would then be able to use the hypermedia server program to explore this repository. For every bit of text they access, the user would pay a royalty to its publisher, even if the particular document they compiled from their session in Xanadu was made up of texts from a variety of different publishers. This royalty payments system was a key component of Xanadu, since it would, Nelson argued, create a market, a trade of texts that would encourage more publishers to contribute, which, in turn, would attract more users.

The network of databases would be set up along the lines of the franchising network of McDonalds, where the infrastructure grows in pace with the market, avoiding the need for costly and risky long-term investment. Companies would provide computer storage for the texts and, in return for the right to use the Xanadu name and the server software, make it available to the network. As more storage is needed so more franchisees would be attracted.

The idea of a fast-information outlet, doing for data what McDonalds did for burgers, is an intriguing one. It is an imaginative attempt to reinvent that most important of social institutions, the library. With a conventional library, you can only borrow work, you cannot, at least not without becoming an author and finding a publisher, contribute to it. When the playwright Joe Orton so famously did in 1962, adding rude comments to books borrowed from his local library, he fetched up in jail. In Xanadu, there is no such sanction. Every 'reader' becomes a potential 'writer', the system being specifically designed to make it as easy to contribute a text (or music, or video, or whatever) as it is to consult one. It creates an open market, a free trade in knowledge, where the success of a text is simply dependent on the number of times it is accessed, which in turn will depend on its relationship with other texts on the system.

In Xanadu even the book is, as Nelson implied it was when he over-eagerly anticipated its demise, an obsolete cultural artefact. There can be no fixed text. Everything within Xanadu exists by virtue of its links with everything else, and those links are constantly being forged and broken. Every reader of every text contributes to its meaning by participating in the creation of the structures that place it. Such links, through indices, quotes, references and notes, are already part of the way we use literature; in Xanadu, they will be more sophisticated. When Xanadu is mentioned in a text about Ted Nelson, there will be no need to scurry off to a poetry collection that may or may not contain Coleridge's poem, 'Kubla Khan', or to a history of the American press so that you can discover that it was the name of William Randolph Hearst's mansion (which in turn may link you to *Citizen Kane*, Orson Welles's film based on the life of Hearst). These and any other references to Xanadu will be part of the constellation of links that surround the text about Ted Nelson, along with, perhaps, annotations about the latest news on the development of the Xanadu server program and clips from a TV chat show in which Nelson talks about his next project. At no moment will the file be closed on this text – though, perhaps, as even Nelson slides into history, some of the links will loosen through neglect, or new ones, connected perhaps to hagiographies about the pioneers of computing, will emerge.

Projects like Xanadu, have succoured a growing conviction in the computing community that, ultimately, computers will introduce a new, liberating order that extends throughout all media. Virtual reality has reinforced this conviction, promising to do to theatre, cinema and TV what Xanadu would do to books. 'Whereas film is used to show a reality to an audience, cyberspace is used to give a virtual body, and a role, to everyone in the audience,' wrote Randal Walser, who led Autodesk's 'Cyberspace' virtual reality research project.[9] Walser came up with a new category of artist, the 'spacemaker', who does not create narratives, rather the cyberspace in which the audience can create a narrative for itself. 'Whereas the playwright and the filmmaker both try to communicate the idea of an experience, the spacemaker tries to communicate the experience itself. The spacemaker sets up a world for an audience to act directly within.'

Walser envisaged the cinema's successor as a 'cyberspace playhouse', a place where 'people go to play roles in simulations'. These playhouses would contain all the equipment – headsets, screens, sensors, 'props to give the player solid analogs of virtual objects' – that is necessary to give the audience the experience of full bodily immersion in a virtual reality.

Something very like cyberspace playhouses were already beginning to crop up in America, Japan and Britain in 1990 and 1991. The BattleTech Center, described by its developers, Virtual World Entertainments, as 'the world's first multiplayer, interactive, real-time simulator allowing people to exist and interact in a "virtual universe"', was installed in Chicago's North Pier in 1990. Players sit in a simulated battle tank cockpit, which they manoeuvre round a virtual space seen through a 25-inch screen (the tank's 'window') using a formidable array of controls. The 'tanks' are networked together, so each player sees the other as a tank moving around the same battlefield. In Britain in 1991 a W Industries 'Virtuality' system was installed in an arcade in London's West End. This allowed a number of players, all wearing headsets and sitting in modules with joysticks embedded in the arm rests, to 'fly' Harrier Jumpjets together.

The promoters of virtual reality have found reassuring parallels between these early forays of their technology into the public domain and the origins of cinema, which also began with attention-grabbing technological spectacles at exhibitions and entertainments centres. Indeed, there are striking similarities between the early efforts of cinema's precursors and the claims of companies like Virtual World Entertainments. At the end of the eighteenth century, the Edinburgh painter Robert Barker built a rotunda in Leicester Square – just a few hundred yards from the site W Industries would be demonstrating its virtuality system two hundred years later – in which he hung painted panoramas of exotic locations to create the 'illusion' of being in the location depicted.[10]

However, drawing a crowd is no guarantee that the technology, at least in the form it is demonstrated, will succeed. No less an innovator as Thomas Edison, inventor of the electric light bulb and the phonograph, came up with the 'Kinetoscope' as a device for displaying moving images, though this missed out the vital element of projection, being modelled on 'What the Butler Saw' machines that could only accommodate one viewer at a time. So it would be premature to predict a mass media revolution on the basis of the popularity of a few arcade games and exhibition demonstrations.

One of the main issues that remains unresolved is how an entire media infrastructure created for mass markets and passive consumption can produce the promised new era of personalized interactive hypermedia.

The Media Lab has led the way in addressing this issue in its prognostications over the future of television – a future it sees as under threat from the international debate to determine the TV broadcasting standard for the twenty-first century. The problem with the debate for the Media Lab's founder, Nicholas Negroponte, was that it was over the wrong issue: high-definition television. 'If you walk down the street and ask anybody "What's wrong with television?"', Nicholas Negroponte told a group of Canadian businessmen in 1989, 'you will not encounter a single person who will answer "resolution". The answer you will get is programming.'[11] So why, Negroponte reasonably asked, all the fuss about high definition television, HDTV? 'If you compare broadcast quality NTSC with broadcast quality HDTV, the difference is minimal', he claimed, stretching a point, perhaps, a little far, particularly for Europeans who cannot for one moment understand how Americans can tolerate the bilious hues of the picture produced by America's NTSC TV broadcasting standard.

At the time Negroponte was speaking, both the Japanese and European were promoting HDTV standards based on analogue technology, which would effectively force the TV set to remain the passive receiving device that it was the dream of the promoters of hypermedia to banish. For Negroponte, HDTV represented 'dead technology' because it was not digital. 'The television is probably the stupidist consumer home electronics product', he said. 'Your refrigerator has more microprocessors.' What we all need, according to the Media Lab, is 'open architecture TV'. This entails turning the TV set into a computer that can process all the digital information that will come into the home – via cable, the telephone network as well as, ultimately, satellite broadcast signals (good old terrestrial broadcasting, with its limited information-carrying capacity, does not have much future in either a digital or HDTV future). It will display this information in accordance with the specific capabilities of the equipment (more expensive sets will be able to deal with more information and display it, should the user wish, in more sparklingly sharp resolution) and the viewer's needs. In such a regime, the old, arbitrary distinctions between print and visual media may gradually disappear, as the new, smart TV set's printer takes over the business of publishing a personalized newspaper each morning, featuring stories taken direct off newsagency wires and electronic mail boxes by an 'agent', a software program that knows its owner's interests and downloads stories that are likely to appeal to them.

According to the Media Lab open architecture scenario, broadcasters will not send 'facsimiles' of TV pictures; they will send the information needed to construct the picture itself. For example, the set for a particular scene in a TV drama could be sent in the form of a three-dimensional geometrical model that the TV receiver reconstructs for itself. The broadcaster then only need send the picture data for the characters and props that populate the set, which the TV integrates with the set. Computer models of film sets might be sold much like real estate – hyperreal estate, as it would inevitably be called – which viewers would then use as the backdrop for different dramas, including, of course, ones in which they themselves feature. In terms of virtual reality, this will mean that TV will ultimately become what could be called an 'immersive' medium. In Walser's language, the front room will be a cyberspace playhouse, the TV set a cyberspace engine. Programming will be built up out of both data sent in over a fibre optic cable network (predicted by the technologists to be the virtually universal form of information distribution by the end of the century) and information held in the TV set itself (perhaps models of the family's personalized virtual bodies), and used to generate worlds for a newly uprooted generation of couch potatoes to explore. Watching TV will become an interactive experience, with the front room's furnishings participating in the creation of the virtual worlds that the viewer temporarily enters. Being absorbed in a programme will take on a more literal meaning, one William Gibson realized in a more conventional, and critical, literary way in a passage from his novel *Count Zero*: 'He checked the time on the kiosk's Coke clock. His mother would be back from Boston by now, had to be, or else she'd miss one of her favourite soaps. New hole in her head . . . she'd been whining for years about static and resolution and sensory bleed-over, so she'd finally swung the credit to go to Boston for some cheapass replacement. . . . He knew her, yeah, how she'd come through the door with a wrapped bottle under her arm, not even take her coat off, just go straight over and jack into the Hitachi, soap her brains out good for six solid hours. Her eyes would unfocus, and sometimes, if it was a really good episode, she'd drool a little.'[12]

The visions of people like Negroponte and Nelson seem to be founded on a belief that mass media have been imposing their ideas on the readers and viewers too long and that new technology will initiate what Stewart Brand dubbed a 'personal renaissance': 'Marshall McLuhan used to remark, "Gutenberg made everybody a reader. Xerox made everybody a publisher." Personal computers are making everybody an author. . . . If, as alleged, the only real freedom of the press is to own one, the fullest realization of the First Amendment is being accomplished by technology, not politics.'[13] In cyberspace, everyone is an author, which means that no one is an author: the distinction upon which it rests, the author distinct from the reader, disappears. Exit author . . .

Notes

1 Jorge Luis Borges, 'The Library of Babel', in *Labyrinths*, London: Penguin, 1970, p. 81.
2 John Walker, 'Through the looking glass', Autodesk internal paper, 1988, p. 6.
3 In John Calder, *A William Burroughs Reader*, London: Picador, p. 19.
4 Levy, 1985, p. 140.
5 Michael Joyce, 'Selfish interaction: subversive texts and the multiple novel', *Jackson, Mississippi: Center for Narrative and Technology, 1988, p. 6.*
6 *Brian McHale, Postmodernist Fiction*, London: Routledge, 1989, p. 109.
7 Michael Swaine, 'Building Xanadu', *Dr Dobbs Journal*, April 1991, 16 (4), p. 111.
8 Lecture to the Institute of Information Scientists at the Geological Society, London, 3 October 1990.
9 Randal Walser, 'Elements of a cyberspace playhouse', *Proceedings of the National Computer Graphics Association* '90.
10 John Wyver, *The Moving Image: an international history of film, television & video*, Oxford: Blackwell, 1989, p. 9.
11 Doug Powell, 'Voice to be primary user interface of '90s, CBTA told', *Computing Canada*, 12 October 1989, 15 (20) p. 74.
12 William Gibson, *Count Zero*, London: Grafton, 1987, p. 54.
13 Brand, 1989, p. 253.

9
FICTION

'Perhaps Hamlet is the first character to stop in his tracks and mutter three miniscule and infinite words that suddenly open a void between the certain truths of the Middle Ages and the uncertain reason of the brave new world of modernity', wrote the Mexican novelist and essayist Carlos Fuentes. 'These words are simply that: "Words, words, words . . . ", and they both shake and spear us because they are the words of the fictional character reflecting on the very substance of his being.'[1] Fuentes wrote this in an introduction to Cervantes's *Don Quixote*, a work often cited as the prototype novel. He was arguing that fiction like *Don Quixote* and *Hamlet* marked a transition from the medieval world of canonical certainty to a modern world of open enquiry, a world where 'all things become possible'. When Don Quixote demands: 'Believe in me!', when he expects us to believe that sheep are armies, inns are castles, windmills are giants to be tilted at, we are shown, Fuentes argues, how fictions cannot leave fact unviolated. 'Reality may laugh or weep on hearing such words. But reality is invaded by them, loses its own defined frontiers, feels itself displaced, transfigured by *another* reality made of words and paper.'

The text may have been old, but Fuentes' analysis of it was thoroughly modern. The notion of there being a 'reality made of words and paper' is an idea truly of our times, times that have seen a sustained assault on the fundamental notion of reality. The culmination of this assault has been the emergence of 'postmodernism', a word that embraces both an intellectual movement and climate, both a way of changing the world and of understanding it. One of its champions, Charles Jencks, describes it as a 'new world view', the view that, unlike its predecessors, is capable of explaining the dominant phenomena of our times: the media age, the global village, catastrophe and chaos, hyperreality and cyberspace. Artificial reality is the authentic postmodern condition, and virtual reality its definitive technological expression.

Every book or meeting on postmodernism begins with attempts to define it. In 1985, at a conference held at the Institute of Contemporary Arts in London, that task fell to Jean-François Lyotard. No one could have been better qualified for the job. First, he is a French intellectual, and follows in his country's tradition of creating sophisticated, elegant, compelling theoretical accounts of key political, cultural and historical issues without recourse to crude matters of fact. Second, he has written a book called *The Postmodern Condition* in which the terminology of postmodern analysis – terms like narrative, metaphor, text and discourse – are much in evidence, even when he is not apparently writing about language. Third, and most important of all, he does not define postmodernism directly, but outlines the arguments that help define it. No one dare confront postmodernism without the reflective shield of argument and analysis.

There are, he said, three debates 'implied by, and implicated in, the term "postmodernism"'.[2] One concerns ways of thinking about art and other forms of expression, which, in postmodern terms, is conducted by a process he compared with Freudian analysis. 'The "post-" of postmodernist does not mean a process of coming back or flashing back, feeding back,' he said, 'but of *ana*-lyzing, *ana*-mnesing, of reflecting.' (Commentators of postmodernism have a fondness for word-play, discovering significance in linguistic coincidences.) The second aspect is a loss of confidence in the idea of progress – 'there is a sort of sorrow in the *Zeitgeist*,' according to Lyotard. 'The development of technosciences has become a means of increasing disease, not of fighting it.' The third aspect was that 'there is no longer a horizon of universalization, of general emancipation before the eyes of postmodern man.' Postmodernism represents a 'rupture' with the modernist belief in universal truths.

However, as Lyotard observed, this third aspect, possibly the one that is most widely identified with postmodernism, is itself very 'modern'. 'Since we are beginning something completely new, we have to re-set the hands of the clock at zero. The idea of modernity is closely bound up with this principle, that it is possible and necessary to break with tradition and to begin a new way of living and thinking.' This time, though, things are different, because 'we can presume that this "breaking" is . . . a manner of forgetting or repressing the past. That's to say repeating it. Not overcoming it'.

Postmodernism, then, represents a break with 'modernism'. So to understand one, we must try to understand the other. And to do this, we could do worse than begin by isolating that moment when the two confronted one another, when the modern met its last post.

The architectural critic Charles Jencks has come up with a satisfyingly specific candidate for such a moment, at least with respect to architecture: 'Happily, we can date the death of Modern Architecture to a precise moment in time', he wrote. 'Unlike the legal death of a person, which is becoming a complex affair of brain waves versus heartbeats, Modern Architecture went out with a bang . . . [It] died in St Louis, Missouri on July 15, 1972 at 3.32pm.'[3] The occasion was the blowing up of the prize-winning Pruitt-Igoe housing scheme . The scheme was classically modern. It was constructed according to the principles of the Congress of International Modern Architects, which put economic and socio-logical issues above those of style as the imperatives of architectural progress.

It is, perhaps, thanks to the way the word 'modern' has acquired such a bad name in architecture that enables us to see most clearly how the term can have a meaning other than the most obvious, how it can come to denote a whole set of assumptions about the way the world works and should be organized. 'It should be pointed out that Modern Architecture is the offshoot of Modern Painting and the Modern Movements in all the arts,' wrote Jencks, in his effort to assure a proper distribution of blame. 'Like rational schooling, rational health and rational design of women's bloomers, it has the faults of an age trying to reinvent itself totally on rational grounds.'[4]

When Fuentes wrote of the 'brave new world' of modernity, he was referring to what seems to be a widely accepted view of history. According to this view, the past can be understood as falling into distinct eras. These eras are not usually associated with particular events, since they are concerned with what one could call the prevailing cultural mood, the dominant ideas and beliefs that characterized the way people thought and behaved at particular times; their assumptions, their attitudes. In this view of history, each era is haunted by its own *Zeitgeist*, the spirit of the times, and it is the task of those who identify and describe such eras to bust the ghost.

Perhaps three great eras can be listed without too much controversy: the Classical era of the ancient Greeks, the medieval era and the modern era. The modern era, from the deep perspective of the historian of human thought, began after the Renaissance, or possibly with the Enlightenment, or at any rate whenever ideas such as technological progress and the discovery of knowledge through reason and science began to undermine the authority of religion. Modern*ism* can be thought of as the self-concious response in the arts to the experience of modernity that appeared at the end of the nineteenth century and the beginning of the twentieth century.

'To be modern', wrote Marshall Berman, 'is to find ourselves in an environment that promises us adventure, power, joy, growth, transformation of ourselves and the world – and, at the same time, that threatens to destroy everything we have, everything we know, everything we are . . . To be modern is to be part of a universe in which, as Marx said, "all that is solid melts into air".'[5] Indeed, Berman used Marx's words as the title of his powerful study of the 'experience of modernity'. Berman's book is not simply an intellectual survey. It concerns very real issues and experiences, a fact emphasized by his dedication of the book to his son, who died shortly after it was completed. 'His life and death bring so many of its ideas and themes close to home,' wrote Berman, 'the idea that those who are most happily at home in the modern world, as he was, may be most vulnerable to the demons that haunt it.'[6] Modernity is not a remote, abstract concept; it is a very real force of very direct impact. Like Mickey Mouse invoking the Sorcerer's spells to increase his productive power, we have conjured up a *Zeitgeist* we cannot control.

The birth of modernism, like its death, is an event that can be traced to one moment, albeit it one created in fiction. Marshall Berman quotes from the prose poem called 'Loss of a halo' written in 1865 by the the French poet Charles Baudelaire, who Berman claims 'did more than anyone in the nineteenth century to make the men and women of his century aware of themselves as moderns'.[7] The poem takes the form of a dialogue between a poet and an 'ordinary man', who run into each other in the sort of place – a brothel, perhaps – neither would normally wish to be found. The poet relates an event that happened to him on his journey there: 'Just now as I was crossing the boulevard in a great hurry, splashing through the mud, in the midst of a moving chaos, with death galloping at me from every side, I made a sudden move, and my halo slipped off my head and fell into the mire of the macadam.'[8] Berman describes this as a 'primal modern scene', that of modern man confronting the 'maelstrom' of the modern city street, forced to live at the pace of the traffic around him.

Sixty years later, the great twentieth century architect Le Corbusier, was walking those same Parisian streets, observing a world 'gone mad' with the 'fury of the traffic'[9] – indeed, being driven off the street by it. Le Corbusier's reaction, however, was very different to Baudelaire's: he suddenly saw 'the titanic rebirth of a new phenomenon'. 'One is seized, filled with enthusiasm, with joy' at being in the 'midst of power, of strength'. It was a revelation so forceful that it seemed to blind him to a recollection he had just had that twenty years earlier the street had 'belonged to us'. He seemed to be possessed by the modernist message, arguing that the street be turned into a 'machine for traffic', just as he argued that the house should become a 'machine for living in'. He saw the road as a factory line, a conveyor belt designed to produce traffic. 'This modern sentiment is a spirit of geometry,' he wrote. 'Exactitude and order are its essential conditions. . . . In the place of individualism and its fevered products, we prefer the commonplace, the everyday, the rule to the exception.'[10]

The slogan of the architectural modernism that Le Corbusier inspired was that form should follow function. In other words, style was nothing, efficiency everything. The design of a building should be dictated by what the building does, so that it does it as well as possible. It should not be an expression of individual creativity but the product of social necessity. Hence the highrise: the efficient, automated, standardized housing system that maximized scarce resources. The Pruitt-Igoe scheme was just such a system. 'It consisted of elegant slab blocks 14 storeys high', wrote Jencks, 'with rational "streets in the air" (which were safe from cars, but as it turned out, not safe from crime); "sun, space and greenery", which Le Corbusier called the "three essential joys of urbanism" (instead of conventional streets, gardens and semi-private space, which he banished) . . . its Purist style, its clean, salubrious hospital metaphor, was meant to instil, by good example, corresponding virtues in the inhabitants.'[11]

The ideals embodied at Pruitt-Igoe were not, as Jencks observed, confined to architecture. Modernism is a movement that expressed itself throughout culture, in response to a world in which the old cardinal certainties were dissolving in a seething cauldron of change. In 1913, Albert Einstein published his *General Theory of Relativity*. It was a full-frontal assault on the tenets of Newtonian physics, the theory that had dominated the modern (as opposed to modernist) era with its view of the universe as a sort of stage upon which objects, lumps of matter, are merely players reacting to one another through the exchange of energy.[12] According to the theory of relativity, the distinction between players and stage could no longer be made – space and time exist, he argued, as much in the interaction of matter and energy as matter and energy exist within space and time.

The theory of relativity, however, was anything but relativistic – you could argue that, by dealing with the increasingly troublesome anomalies resulting from attempts to apply Newtonian principles to newly observed phenomena, it actually saved science from relativism. But it did deal with the central struggle of the modern era: finding a way of establishing universal truths in a world yielding many different, constantly shifting points of view. In this respect, the theory of relativity can be seen as a prototypical modernist, as opposed to just modern, theory, because it was a response to the disorientating experience of modernity.

The mathematician David Hilbert, as we have seen, marked the turn of the twentieth century with a speech given to the International Congress of Mathematicians in Paris. The modern era had subjected the mathematical system, based on principles reaching back to the classical era of the ancient Greeks, to considerable stress, and Hilbert enumerated the problems, many of them fundamental, that this had produced. As in physics, a new theoretical framework was required, one that meant maths was no longer grounded in the shifting sands of physical reality, but in some higher, abstract, formal realm.

Modernist art addressed this shift to abstractionism with ideas such as cubism, which attempted to depict one scene simultaneously from many different points of view. Picasso saw the First World War as a cubist war, since it was fought on so many fronts.[13]

Trouble had been brewing in the visual arts since the invention of photography. Deprived of its monopoly on depiction, painting needed new ways of justifying itself. In his 1859 essay 'The modern public and photography', Baudelaire railed that 'where one should see nothing but Beauty (I mean in a beautiful painting) our public looks only for Truth'.[14] This is a revealing phrase. It is clearly no acceptance of defeat, allowing that photography be accepted as a sign of artistic progress. It asserts art's unique access to beauty as unchallenged. Photography is used as evidence that art is about more than truth, at least the obvious sort yielded by the physical world; that it is about a higher truth or a higher reality, or at any rate something better than what we get at ground level.

This reaction, though expressed by Baudelaire in a typically bad-tempered tone, and prompted by his growing disgust with modernity, was, perversely, an essentially progressive one. It reflected art's return to the wild, its forced liberation from the realistic depiction of the physical world – just as the development of new mathematical techniques liberated maths from the assumption that its validity came from how closely it modelled the physical world. It was not simply that the modern world was becoming too hot to handle, too difficult for a single, unitary system or set of values to embrace. It was that art, like maths, had gone beyond the mundane fact of physical reality in search of something more exciting. As Picasso later put it: 'We all know that Art is not truth. Art is a lie that makes us realize the truth, at least the truth that is given us to understand.'[15]

At the beginning of the twentieth century, painters began to experiment with their new-found freedom. One of the most experimental was the Russian artist Kazimir Malevich, who formalized his discoveries in a new artistic movement he called 'suprematism'. Despite the fascist sounding name, suprematism was concerned with aesthetic, not racial, ascendency. It claimed to offer a sort of super-realism, overcoming 'the terrestrial pull' of naive realism.

The best-known suprematist work is Malevich's *Black Square*, painted in 1914 and 1915. It is no more than its title says it is: a black square, on a white background. The only features on *Black Square* other than a black square are cracks in the black paint, which are the work of age rather than Malevich. This is the sort of art that, even at the end of the century it helped to introduce, people outside the art world find hard to accept has having any aesthetic merit. Its abstraction, its wilful lack of representation, is almost as alien and difficult now as it was when it was first exhibited in 1915, even after Mark Rothko's bigger black squares, and Mies van der Rohe's black square buildings.

'The keys of Suprematism,' wrote Malevich,[16] 'are leading me to discover things still outside of cognition. My new painting does not belong solely to the earth. The earth has been abandoned like a house, it has been decimated. Indeed, man feels a great yearning for space, a gravitation to "break free from the globe of the earth".'

What sort of space was man yearning for? When the Renaissance artist Filippo Brunelleschi began making experimental drawings with mirrors and callipers, he was attempting to find a scientific way of projecting a three-dimensional space onto a two-dimensional surface, the canvas. This was a fundamentally modern enterprise: it attempted to apply rational, scientific principles to the act of depiction. Early perspective painting was so in the thrall of this new mechanism for capturing reality that it demanded an awesome array of analytical apparatus. The artist seemed better prepared to produce a surveyor's report than a work of art.

However, by Malevich's time, the space projected by perspective was seen as an imprisonment, not a liberation. The underlying principle of perspective was that all points of view are interchangeable; it was a system that could compute what the same objective reality would look like from different positions. But no such computation seemed to work in the contemplation of the fracturing structure of modern experience. Perspective became a limitation on what the artist could depict, not an expansion of what the picture could show. 'Any sight is a sum of different glimpses', wrote the art critic Robert Hughes. 'And so reality includes the painter's efforts to perceive it. Both the viewer and the view are part of the same field. Reality, in short, is interaction.'[17]

The German philosopher Friedrich Nietzsche is, perhaps, the archetype hero of modernism; indeed, one might romantically think that it was the contemplation of the uncertainties of the modern experience that drove him mad. He is credited with throwing rationalism and truth into the maelstrom of modernity – not even these, he argued, could be elevated above the battlefield of life. It was from this raging intellect that the theory called perspectivism is first said to have emerged: the idea that there is no way of establishing independently one perspective on the world as more valid than another. There are as many realities as points of view, as Spanish philosopher José Ortega y Gasset, who gave the theory its name and formal description, put it. There is only one reality, but that is the reality of each individual's life. This idea he expressed in the nostrum: 'I am I and my circumstances' – 'I' only exist in interaction with the environment in which 'I' live – and he expressed it in his first major book, a study of the text that had inspired Fuentes: *Don Quixote*.[18]

It was such ideas that expelled authors from their books, an act sanctified in the slogans of modernist critics such as Joseph Warren Beach, who coined the term 'exit author' in 1932 as a challenge to the Victorian writers' lofty assumption of a privileged perspective. Modernist writers like James Joyce, as modernist architects like Le Corbusier, heroically denied themselves their traditional authorial role, instead trying to hide themselves behind 'narratorless' texts and structures yielded by the mechanical application of rational writing and architectural systems rather than divine inspiration (a strategy that resulted in self-publicity rather than self-abnegation, with modernist artists achieving a celebrity that even their predecessors would have envied – the author may have exited, but he made one hell of a racket backstage). The challenging, fractured narrative of modernist fiction attempted to break the spell of the single, linear plot. Bertolt Brecht's plays wilfully set about discomfiting – 'estranging' – their audiences, tugging their disbelief back to the stage floor before it manages to escape into suspension, in order to show the historical, transient nature of the events represented.

As Baudelaire anticipated, art was to respond to the public's search only for truth in the material world by searching for it elswhere. Modernism was about disenchantment with material truth and the search for abstract truth. It developed its own self-destructive dynamic to show how the physical world was being eaten up by the constant need of capitalism to renew itself, stoke up growth, keep the machine moving, replace the old with the all new, as Marxist critics like Marshall Berman see it. 'The innate dynamisn of the modern economy and of the culture that grows from this economy,' wrote Berman, 'annihilates everything that it creates – physical environments, social institutions, metaphysical ideas, artistic visions, moral values – in order to create more, to go on endlessly creating the world anew.'[19] And so it was that the Pruitt-Igoe, and with it modernism, was itself annihilated.

What has replaced it, according to Jencks, is another world view, the one unhelpfully called postmodernism – a trend whose termination, as Jencks reported in the introduction to the sixth edition of his key text (insofar as postmodernism allows such a thing as a key text) *The Language of Post-Modern Architecture*, was itself announced by its detractors in 1982. A group of modernist stalwarts marked the event with doctored photographs of Michael Graves's postmodern Portland Building showing it suffering the same fate as the Pruitt-Igoe. To the champion of postmodernism, such an announcement was, of course, premature – even evidence of its success: 'Now that Post-Modern architecture has triumphed around the world, many people have declared it dead', Jencks wrote. 'This, the fate of all successful movements, is something to be celebrated. Born in a fit of love, they grow to maturity all too quickly, are vulgarized, mass-produced and finally assigned to the scrap-heap of history.'[20]

When Prince Charles announced that the Mies van der Rohe office tower intended for the City of London was a 'glass stump', he was not articulating a postmodern perspective – Jencks calls it 'premodernist'. Nevertheless, just as the persistence of poverty and alienation had discredited the tower block and large-scale public development scheme as the instruments of postwar social salvation, so the prince had lobbed an attractively weathered brick through the plate glass of the modernist aesthetic. No more towers, no more technology, no more steel or stone – it was fitting that a set of ideas originating in revolutions that toppled thrones should be dispatched by an heir to one.

In architectural terms, postmodernism has replaced the universalizing, standardizing, unifying, rationalizing, simplifying instincts of modernism with pluralism and complexity.

Postmodernism is the ism that isn't. It is about the accommodation of lots of different isms. 'Fundamentally,' wrote Jencks, 'it is the growing understanding that pluralism creates meaning; or put negatively in the cool terms of information theory, that "where there is no difference, there is no information".'[21]

* * *

In 1982, the American academic Frederic Jameson presented a lecture at the Whitney Museum on postmodernism. In 1983, this formed part of a paper entitled 'Postmodernism and consumer society', which was published in a number of collections.[22] In 1984, he published another paper in the *New Left Review* entitled 'Postmodernism: the cultural logic of late capitalism',[23] which in 1988 was incorporated with the original 'Postmodernism and consumer society' paper to create a new paper, printed with the latter's title in another collection.[24] Finally, in 1988, this last version of the paper, still carrying traces of the Whitney Museum Lecture, was used as the basis of the first chapter of a book published in 1991 entitled *Postmodernism, or, the cultural logic of late capitalism* – the same title as the 1984 paper, only with an 'or' added.

The core ideas of the works remained roughly the same throughout this evolutionary development, with individual examples coming and going according to the necessities of fashion. For example, Michael Herr's book on the experience of Vietnam, *Dispatches*, appeared in the 1988 version, but was dropped from the 1991 book. In the 1983 paper, schizophrenia was used as a metaphor for the postmodern condition. It disappeared in 1988, only to reappear – virtually untouched – in 1991. If ever there was a demonstration of the redundancy of the idea of a finished text, a confirmation of Ted Nelson's concept of the hypertext, this was it. Never exactly the same words, never exactly in the same order, but nevertheless the same basic ideas – a 'living' document, kept alive by its constant turnover of material, discarding dead and dying bits as fresh new ideas and examples burst forth.

Jameson himself seems to be a perfect example of the postmodern condition. Not only do his texts stay in a constant state of flux reflecting the pace of cultural change around them, but he is himself a product of an academic world where the fixed, modern categories of academic study are under assault. At the time of

publishing the first version of the paper he was 'Professor of Literature and History of Consciousness at the University of California,
Santa Cruz'. Such formidable titles are becoming a common
feature of the American campus. The modern era's classifications
of knowledge into the humanities and sciences are collapsing.
This is something Jameson himself regards as one of the
symptoms of postmodernity. 'A generation ago', he wrote in the
1988 version of his paper, 'there was still a technical discourse of
professional philosophy – the great systems of Sartre or the phe-
nomenologists, the work of Wittgenstein or analytical or common
language philosophy – alongside which one could still distinguish
that quite different discourse of the other academic disciplines –
of political science, for example, or sociology or literary criticism.'[26]
Now, he argued, the 'discourses' are dissolving into a new kind of
writing called simply 'theory'. 'This new kind of discourse,
generally associated with France and so-called French theory, is
becoming widespread and marks the end of philosophy as such.
Is the world of Michel Foucault, for example, to be called
philosophy, history, social theory or political science? It's undecidable, as they say nowadays; and I will suggest that such
"theoretical discourse" is also to be numbered among the
manifestations of postmodernism.'

Jameson's announcement of philosophy's death is greatly
exaggerated, certainly with respect to British universities, where
'theory' is treated with suspicion where it is not simply
dismissed. The body of postmodernist theoretical literature
remains largely hidden by the absence of any widely accepted
category to place it in. For example, any of the collections that
contain Jameson's work cannot be easily placed in any of the
conventional library or bookshop sections: is it literary criticism,
philosophy, sociology? I found *Postmodernism and its Discontents*
in the media studies section in one well-known London shop,
under literary criticism in another. The Dewey cataloguing
system developed by American librarian Melvil Dewey at the end
of the nineteenth century – itself one of the great monuments of
modernity, with its heroic attempt to order the sum of knowledge
with a neat numbering system – is already stretched to breaking
point, having to accommodate subjects that were never anticipated
when it was first devised. The system has been unable to find a
number that fits 'theory'.

What, then, is 'theory'? It is possibly better described as 'critical theory', because the feature that makes it distinct from other forms of theoretical study is the concept of 'critical' investigation. The word 'critical' has had a special resonance in German philosophy since Immanuel Kant wrote his *Critique of Pure Reason*, which developed the concept of 'critical idealism'. In 1923, the University of Frankfurt founded the Institute for Social Research, which later became known as the 'Frankfurt School', with a brief to create a 'critical theory' of Marxism. The critical approach was contrasted with 'positivism', which attempted to explain political and social phenomena using the same, value-free, objective strategy of sciences such as physics. A critical approach to philosophy and politics could be seen as having parallels with a critical approach to art. That is why the most important works of the members of the Frankfurt school and those that followed them appear to be works of literary criticism. For example, Walter Benjamin, perhaps the school's most influential member, did so much to reveal the nature of modernism through a study of Baudelaire.[27]

While critical theory has its origins in Germanic culture, the other main component of 'theory', structuralism, came from the French-speaking world, in particular from the work of Swiss linguist Ferdinand de Saussure. Saussure saw language as a system of signs in which words (the signifier) have only an arbitrary relation to what they signify (the word's meaning). A tree has no special link with the sound of the word 'tree'. A word, therefore, gets its meaning from the way different signs relate to each other – just as, to use one of Saussure's own analogies, the 'meaning' of a move in chess only arises out of its context within a game.

Structuralism provided a basis for using a new set of techniques not just to examine language but, by extension, the way we make sense of the world as a whole. The French anthropologist Claude Lévi-Strauss argued that language constitutes 'at once the prototype of the *cultural phenomenon* (distinguishing man from animals) and the phenomenon whereby all the forms of social life are established and perpetuated.'[28] And he went further. He sought to discover 'whether the different aspects of social life (including even art and religion) cannot only be studied by the methods of and with the help of concepts similar to those employed in linguistics, but also whether they do not constitute phenomena whose inmost nature is the same as that of language.'[29] His answer to both questions was an emphatic yes, and the means he used to demonstrate it was to examine the structures that exist in social 'systems' such as kinship, the sets of relationships that underpin the family unit.

Together with critical theory, structural anthropology implied that the modern world could be better understood as a vast text that could be critically examined using the same basic techniques as one might examine a novel or play. Furthermore, this could be done without the critic having to dislodge him- or herself from the university library, since the text itself was to be found in society's cultural output – its books, its plays, its films, its television, its newspapers. In Lévi-Strauss's most famous work, *Tristes Tropiques*, an intellectual authobiography recalling his studies of central Brazilian societies, he showed a distaste for fieldwork, which was a good thing, since what little he did was regarded by other anthropologists as sloppy – he did not learn local languages or spend much time with his subjects. From the point of view of the empiricist tradition of Anglo-Saxon, and in particular British, culture, nothing could better commend consigning a theoretical system to the intellectual scrapheap. From a continental perspective, nothing could assure such a system a more secure basis for fruitful academic enterprise.

And so it is when one reads Frederic Jameson's work, and any other critical analysis of modernism and postmodernism, one finds not a jot of empirical data to support its arguments, any more than one would find Danish patricide statistics used to judge the validity of *Hamlet*.

Where the issue is controversial, of course, is determining the status of such techniques when it comes to judging matters of 'fact' rather than fiction. In 'theory', the distinction between the two is barely visible – in fact, it is part of the condition of postmodernity that there is seen to be no essential difference between the two, at least not in the sense that one is 'real', the other not. According to what might be called the antitheory – or, perhaps, 'practice' – of empirical science, there can be no proper account of the products of human creativity. This obviously presents a difficulty when it comes to dealing with culture. The arts are the product of the imagination, so they are out. The news media passively communicate information from one point to another – though not always accurately, which means they are the product of journalistic imagination; either way, they are out, too. Architecture serves some practical purpose, which means its study is a matter of finding efficient mechanisms for realizing the client's wishes – the client's wishes, however, remain the unexplained product of that inscrutable imagination. Even technology is just 'there'; science may be able to explain how it works, but as to why this particular technology is produced now is a matter of the blind forces of progress (it could not have been invented before because there was insufficient knowledge) and practicality (it is useful). And so on, and so on. At every turn, the phenomena that dominate our lives are resolved into matters of subjective opinion, quite beyond the scope of science to account for. Of course, social sciences such as the sort of sociology and anthropology that tends to be taught in British universities have attempted to use empirical means to analyse social experience, and have had some success explaining the mechanisms of power and control. But they, too, have been hampered by their need to base all their findings on measurable phenomena and testable hypotheses – and have been increasingly excluded from the scientific mainstream for their efforts.

The strength of 'theory', then, lies in what the empiricists regard as its fundamental weakness; its dependence on a 'reading' of the world we live in. Its success is measured not in terms of its correspondence with some objective, eternal 'truth' but according to how well it performs in the great world market of ideas, how well it generates other readings, how much interest it arouses. Everything, even itself, is a story, and its validity rests on how good a story it is.

'Theory' does not even exclude science from this story. Indeed, Lyotard's *The Postmodern Condition: a report on knowledge* (the subtitle – difficult to accept for a book a hundred pages long – comes from it having been commissioned by the Quebec government's Conseil des Universités) sets out to show how the authority of science rests on stories or, in the more technical language employed by critical theorists, how it is 'legitimated' by 'narratives'. Lyotard's ideas are not easy to understand – some even harder to accept – but at their centre seems to lie the notion that science must ultimately rely for its authority on the potency of the myths, or 'metanarratives', that underlie it, and it is Lyotard's task to isolate and examine these myths. He isolates two: one political, the other philosophical. The first is the heroic human struggle for liberty. 'If the social subject is not already the subject of scientific knowledge', he writes, 'it is because that has been forbidden by priests and tyrants.'[30] As Frederic Jameson points out in his introduction to the English version of the book, this is the narrative born of the French Revolution, and its potency is reflected in the fact that we take it for granted that everyone has a right to knowledge, a right reflected in the unquestioning duty of the state to educate the public. While the first narrative might be thought of as French in its origins, the second is Germanic, and, like much German philosophy, exasperatingly opaque. Whatever else it may be, this narrative seems to rest on knowledge's role in the evolution of a self-conscious will capable of making decisions about what it can and should do.

One might argue with his choice of narratives, but the importance of Lyotard's ideas rests in the idea that it is narrative itself that ultimately justifies what science does – an idea that, Lyotard fully appreciates, most scientists would find abhorrent: 'Scientific knowledge cannot know and make known that it is the true knowledge without resorting to the other, narrative kind of knowledge, which from its point of view is no knowledge at all. [But] without such recourse it would be in the position of presupposing its own validity and would be stooping to what it condemns: begging the question, proceeding on prejudice.'[31] In the twentieth century, science has needed legitimation as never before, to justify the enormous expense of the great scientific projects such as space exploration and to earn its sponsors' tolerance of discoveries, such as atomic fission, that many regard as malign. And just as it has needed legitimation, so legitimation has been harder to find. Science has had to leave behind the 'metaphysical search for a first proof or transcendental authority', as Lyotard put it. Hilbert's questions and Gödel's theorem, as we have seen, demonstrated how science rests on a mathematical system that is ultimately just some sort of game, one that can have no resting place, no 'legitimation', in its correspondence with physical reality nor in some pure set of logical principles. Also, Lyotard argues that there is a 'renewed dignity for narrative (popular) cultures'.[32] He sees the resurgence of narrative as part of the 'liberation of the bourgeois classes from the traditional authorities'. In other words, after being discredited by science as a subjective, inaccurate way of representing reality, the story has made a comeback. Perhaps mass media journalism is a demonstration of that, since its whole claim to being able to report what is going on rests on its facility for telling stories. Any journalist is (or should be) aware of this. Whatever he or she is witnessing, it is nothing without its narrative. 'What's the angle?' 'What's the story?'

What is true of science is true for knowledge, politics, society, anything you choose. '"Doing science"', wrote Jameson, 'involves its own kind of legitimation (why is it that our students do not do laboratory work in alchemy? . . .) and may therefore be investigated as a subject of the vaster political problem of the legitimation of a whole social order . . . Doing "normal" science and participating in lawful and orderly social reproduction are two phenomena – better still, two *mysteries* – that ought to be able to illuminate one another.'[33]

The idea that science is a sort of text is by no means a new one – as already observed, Galileo saw it as the product of the 'language of nature', mathematics. Nor is it something with which scientists who concern themselves with the issue – not many practising ones, it has to be said – necessarily disagree. Niels Bohr and Werner Heisenberg, two of the most influential physicists of this century, didn't. 'What we observe is not nature itself', argued Heisenberg, 'but nature exposed to our method of questioning. Our scientific work in physics consists in asking questions about nature in the language we possess and trying to get an answer from experiment by the means that are at our disposal.'[34] What threatens and upsets them is the idea that they are not engaged in a properly objective enterprise – that a scientific discovery is not something that anyone can, given the right apparatus and language, reproduce for themselves.

In his landmark book *The Structure of Scientific Revolutions*, Thomas Kuhn, the MIT professor of the history and philosophy of science, seemed to issue just such a threat. Despite its title, *Structure* is not really a structuralist text in the same tradition of Lévi-Strauss. The book is best known for giving the word 'paradigm' a highly attractive and very useful new meaning. In 1947, so Kuhn's story goes, he was asked to present a paper on the history of science to some students. Leafing through Aristotle's *Physics*, he wondered how a philosopher as brilliant as Aristotle could be so *wrong*. Then came the revelation: Aristotle was not wrong, he was using a different set of concepts and practices, a different theoretical framework, and within that framework he was perfectly right. The difference between Aristotle and Newton was that they did not share the same *paradigm*. One can often see the operation of a paradigm different to ours in the assumptions of writers from another era. In his *Essays* the sixteenth century French writer Montaigne wrote: '"A strong imagination brings on the event," say the scholars. I am one of those who are very much affected by the imagination. Everyone feels its impact, but some are knocked over by it . . . the sight of another's anguish gives me real pain.' Such sentiments make sense to us even now, but not the conclusion he draws from them: 'As I observe a disease, so I catch it and give it lodging in myself. It is no surprise to me that the imagination should bring fevers and death to those who allow it free play and encourage it.'[35] There is a perfectly solid internal logic to this: sickness is infectious in the same way that laughter can be, because sickness is not, according to the medieval paradigm, symptomatic of a physiological state so much as a spiritual one.

What disturbed the scientific world's self-assurance about Kuhn's idea was his assertion that no paradigm can claim universal validity – the Newtonian paradigm was not an improvement over the Aristotelian one, based on the better or more accurate observation of the world. In fact, paradigms are not even comparable, since there is no overarching paradigm, no common standard, by which they can be compared. They are 'incommensurate'. Each is relevant to its own time, adopted by the intellectual elite of the day until the accumulation of what Kuhn called 'anomalies' make it unsustainable, whereupon a new theoretical framework asserts itself in a 'paradigm shift' (an equally famous term, though not originally Kuhn's).

Unfortunately for Kuhn, his ideas were to become bracketed with those of the philosopher Paul Feyerabend. 'Anarchism,' wrote Feyerabend in his famous 1975 essay *Against Method*, 'while perhaps not the most attractive *political philosophy*, is certainly excellent medicine for *epistemology*, and for the *philosophy of science*.'[36] [Italics all *his*.] Feyerabend's aim was to attack the idea of science having a 'method' based on rationalism which offered privileged access to knowledge. 'It is clear . . . that the idea of a fixed method, or of a fixed theory of rationality, rests on too naive a view of man and his social surroundings. To those who look at the rich material provided by history, and who are not intent on impoverishing it in order to please their lower instincts, their craving for intellectual security in the form of clarity, precision, "objectivity", "truth", it will become clear that there is only *one* principle that can be defended under *all* circumstances and in *all* stages of human development. It is the principle: *anything goes.*'[37] This was a message few even within shouting distance of orthodox science could stomach. 'If this all sounds like the grumblings of a failed scientist to you', John Casti tetchily observed in his book on scientific theory and method *Paradigms Lost*, 'it's perhaps worth noting that Feyerabend did at one time study physics and astronomy.'[38]

For Kuhn, science is about anything but anything going, and he became increasingly uncomfortable with way his ideas were being linked with the 'alternative' and anti-science movements that, fuelled by Feyerabend's ideas, emerged in the 1970s as a reaction to the Vietnam War and atomic weapons. He was not aiming to undermine the authority of science. Paradigms, according to a profile of him published in *Scientific American*, provide the 'secure foundations needed for scientists to organize the chaos of experience and to solve ever more complex puzzles. It is the conservatism of science, its rigid adherence to paradigms . . . that enables it to produce "the greatest and most original bursts of creativity" of any human enterprise.'[39]

Nevertheless, what Kuhn comprehensively succeeded in doing was unsettling a strongly held and rarely questioned conviction among scientists that they alone could reveal reality, because they alone had developed a set of techniques that were objective.

* * *

Here is a little narrative. I was once engaged in an argument over the nature of reality with Andrew Hodges, author of an excellent biography of Alan Turing and a mathematician at the time engaged in research work with Roger Penrose, author of *The Emperor's New Mind*. At one moment in our debate, Hodges suddenly stopped contesting the issue. It seemed to him that all we were talking about was words, and, as he economically put it, 'words don't matter.' Not an argument calculated, even by a mathematician, to gain a writer's sympathy. In the common-sense world of science and especially mathematics, 'words' really do not matter, because they represent no reality, they float above the surface of it. In the world of 'theory', however, the very opposite is true. Reality exists in language, in history, in culture, in all the contingencies of human action and creativity, in the very substance of Hamlet's being, in 'words, words, words'. A universal, objective reality, and the science that seeks to reveal it, was the great sustaining myth of the modern age. The question is, then, what happens to it when that myth is punctured, when the paradigm has been shifted? What is left of *reality* in the rubble of the Pruitt-Igoe?

Notes

1   Carlos Fuentes, 'Introduction', in Miguel de Cervantes, *Don Quixote de la Mancha*, trans. Tobias Smollett, London: André Deutsch, 1986, p. xvi.

2   Jean-François Lyotard, 'Defining the postmodern', in Lisa Appignanesi (ed.) *Postmodernism: ICA Documents*, London: Free Association Books, 1989, p. 7.

3   Charles Jencks, *The Language of Post-Modern Architecture*, 6th edition, London: Academy Editions, 1991, p. 23.

4   Jencks, 1991, p. 24.

5   Marshall Berman, *All That Is Solid Melts Into Air: the experience of modernity*, London: Verso, 1983, p. 15.

6   Berman, 1983, p. 14.

7   Berman, 1983, p. 132.

8   Berman, 1983, p. 156.

9   Le Corbusier, *The City of Tomorrow*, trans. Frederick Etchells, Cambridge, Mass.: MIT, 1971, p. 3.

10 Le Corbusier, 1971, p. 38.

11 Jencks, 1991, p. 24.

12 I have borrowed the metaphor from Geoffrey Sampson's excellent article on Einstein in Elizabeth Devine, Michael Held, James Vinson, George Walsh (eds) *Thinkers of the Twentieth Century*, London: Firethorn, 1986, p. 147.

13 David Harvey, *The Condition of Postmodernity*, Oxford: Basil Blackwell, 1989, p. 278.

14	In Berman, 1983, p. 140.

15	Dore Ashton, 'Two statements by Picasso', *Picasso on Art*, 1972 in *The Oxford Dictionary of Modern Quotations*, Oxford: Oxford University Press, 1991.

16	K. Malevich, 'Suprematizm. 34 risunka', Vitebsk: Unovis, 1920, p. 1 In Dmitrii Sarabianov *Kazimir Malevich and His Art, 1900–1930*, trans. Dr John E. Bowlt in *Malevich*, catalogue for exhibition at the Stedelijk Museum, Amsterdam, 1989.

17	Robert Hughes, *The Shock of the New*, London: BBC, 1980.

18	José Ortega y Gasset, *Meditaciones del Quijote*, Madrid: Ediciones de la Residencia de Estudiantes, 1914.

19	Berman, 1983, p. 288.

20	Jencks, 1991, p. 9.

21	Jencks, 1991, p. 10.

22	For example, Hal Foster (ed.) *Postmodern Culture*, London: Pluto, 1983, p. 111.

23	Frederic Jameson, 'Postmodernism: the cultural logic of late capitalism', *New Left Review*, 146, July–August 1984.

24	E. Ann Kaplan (ed.) *Postmodernism and its Discontents: theories, practices*, London: Verso, 1988.

25	Michael Herr, *Dispatches*, New York: Knopf, 1977.

26	Frederic Jameson, 'Postmodernism and consumer society', ibid., p. 14.

27	Walter Benjamin, *Charles Baudelaire: a lyric poet in the era of high capitalism*, trans. Harry Zohn, London: Verso, 1983.

28	Claude Lévi-Strauss, *Structural Anthropology*, London: Penguin, 1972, pp. 358–9.

29	Lévi-Strauss, 1972, p. 62.

30	Jean-François Lyotard, *The Postmodern Condition: a report on knowledge*, trans. Geoff Bennington and Brian Massumi, Manchester: Manchester University Press, 1986, p. 31.

31	Ibid., p. 29.

32	Ibid., p. 30.

33	Ibid., p. viii.

34	Werner Heisenberg, *Physics and Philosophy: encounters and conversations*, New York: Harper, 1958, p. 58.

35	Michel de Montaigne, *Essays* trans. J. M. Cohen, London: Penguin, 1958, p. 37.

36	Paul Feyerabend, *Against Method*, London: Verso, 1978, p. 17.

37	Feyerabend, 1978, p. 28.

38	John L. Casti, *Paradigms Lost: images of man in the mirror of science*, London: Scribners, 1990, p. 38.

39	John Horgan, 'Reluctant revolutionary', *Scientific American*, 264 (5) May 1991, p. 15.

10
HYPERREALITY

'The Gulf War will not take place.' This was the title of an article by Jean Baudrillard published in the French newspaper *Libération* just before the war's outbreak.[1] Although, as French scholars will know, the phrase was an allusion to Jean Giraudoux's play *The Trojan War Will Not Take Place*, Baudrillard meant it, even though Iraqi forces were occupying Kuwait, Western forces were massing on the Saudi Arabian border and ultimata were being dismissed by Saddam Hussein faster than President Bush was issuing them. From anyone else, anywhere else, such a piece published at such a moment would at the very least be regarded as foolhardy, perhaps the sort of forgivable lapse of judgement that one might expect from an academic who has accidentally strayed into the path of current world events. But this was Baudrillard, intellectual exhibitionist par excellence and hero philosopher of postmodernism. For such a man, the onset of hostilities that almost immediately followed his article, far from disproving his assertion, actually confirmed it. The following month, while the battle raged, he wrote a follow-up article entitled 'Is the Gulf War really taking place?',[2] and a few weeks after the hostilities had ended, yet another article, which, together with the first two, he published in a collection entitled *The Gulf War Did Not Take Place*.[3]

The night the Gulf War That Did Not Take Place began to take place, there was a firework display over Baghdad. 'It looks like a Fourth of July display at the Washington Monument', shouted CNN's John Holliman in the network's famous live broadcast, his words punctuated by the crump of smart bombs and the crackle of ack-ack. 'We just heard . . . Holy Cow! That was a large airburst we saw.' The following night the *Sun* newspaper announced: 'Israel hit by Iraq death gas bombs'.[4] CNN reporters donned gasmasks while reporting on the event live from Jerusalem. The following Sunday, the newspapers tried to make sense of the pictures that have assaulted the senses in the previous week. 'Networks call shots in Saddam soap opera', reported the *Observer*. Sue Masterman wrote on the same page: 'This is the war – brought live straight into your living room. The biggest computer game of all time fought out right under your nose.'[5] 'The exact moment a 2,000 lb bomb disappeared down the ventilator shaft at the Air Ministry in Baghdad was caught perfectly on the video in the nose of a F-117A stealth fighter-bomber,' wrote correspondents for the *Independent on Sunday*, 'and was soon available for replay on television. For 27 hours, until the first Scud missiles landed in Israel, this war was like a computer game.'[6] The *Guardian* carried a selection of quotes taken from the week's coverage: 'I feel like a young athlete after his first football match,' said an American fighter pilot. Another US pilot: 'Baghdad was lit up like a Christmas tree. It was tremendous! I haven't seen anything like it since a 4th of July party years and years ago!' A third pilot gave what must be the definitive assessment of the war that did not take place: 'It was exactly like the movies!'

'On the battlefield of the future', General William C. Westmoreland, the US's Chief of Staff, told the Association of the United States Army in 1969, 'enemy forces will be located, tracked and targeted almost instantaneously through the use of data-link, computer-assisted intelligence evaluation and automated fire control.' These words were quoted by Frank Barnaby in his book on the future of war, *The Automated Battlefield*.[7] 'We will see that, given new and foreseeable military technological advances, all forms of warfare – conventional and nuclear, tactical and strategic – between advanced countries will become increasingly automated', Barnaby added. 'There is no technological reason why warfare should not eventually become completely automated, fought with machines and computerized missiles with no direct human intervention. As the battlefield becomes more automated, the battle itself becomes more like a war game.'[8]

Computers are heavily implicated in this process of automation. Indeed, war, computing and virtual reality are tightly interconnected, one shaping the other, all impossible as they are currently executed and understood without the other. As we have seen, it was the pressure to find ways of calculating ballistic tables that forced the pace of the development of the ENIAC, the first electronic computer. It was the need to train increasing numbers of pilots during the Second World War that produced the first electronic flight simulators. It was a US Navy defence contract that laid the foundations of computer simulation and real-time interactive systems through the Whirlwind project. It was work on the head-mounted display technology undertaken at the Wright-Patterson Air Force Base that inspired the first virtual reality experiments at the NASA Human Factors Research Division.

And so the collaboration continues. BBN Systems and Technologies developed SIMNET, a way of linking networks of simulators together so that exercises could be conducted on virtual battlefields.[9] Over a SIMNET network, tank and aircraft simulators scattered across the globe can perform manoeuvres with each other. In the computer-generated battlefield displayed on the simulator screen, other tanks and aircraft that appear are 'driven' by other crews in other simulators, the data on their movements and actions passed along the network so that all the simulated tanks and planes seem to be sharing the same space. The W-Industries 'Virtuality' virtual reality arcade game achieved the same basic effect – reflecting the close connection between automated war and video games.

This collaboration of computing and war reached its apotheosis in the Gulf. 'Smart bombs' – 'bombs with brains', but no souls – became the heroes of the new technological front, though the types that were most heavily deployed were the dumb sort, disgorged from the bellies of B52s on Iraq's entrenched troops. BBC Television's correspondent in Baghdad, John Simpson, famously reported seeing a Cruise missile fly past his hotel window. 'This is a new era of warfare', announced Senator Sam Nunn, chairman of the Senate Armed Services Committee – 'electronic' warfare, the first field-test of equipment developed for the superpower confrontation that never happened.

Command, control, communications – not what is lost in the noise of modern battle, but what generals call the system that lies at its heart. C^3, 'C-cubed', is the technology that enables generals to monitor and manage the battle's conduct from a bunker. During the Gulf War, information was available at any level of detail at all times: from satellites, from AWACS, from reconnaissance aircraft, from radar, from drones, from ground sensors, from intelligence, from maps, from troops. The job of the C^3 systems was to provide a 'picture' of what was going on, so that General Norman Schwarzkopf could, from the depths of his Saudi control centre, issue appropriate orders.

This same picture could almost have been reproduced by the computers of the Pentagon Joint War Games Agency – it could be a military or computing machine at the other end of the wires leading into the bunker. For Schwarzkopf, aspects of the Gulf War's conduct, at least while it was underway, could have been a simulation. So it could have been for many of its participants: missile targets were not real locations but map coordinates displayed on a VDU, troop movements were formations of pixels in computer-enhanced, false-colour satellite images. From a postmodern perspective, the entire war, at least at the level where anyone could make sense of it, was just a pattern on a screen.

According to the French philosopher Paul Virilio, D. W. Griffith, the only film director allowed to film at the front during the First World War, declared that he was 'very disappointed with the reality of the battlefield'.[10] It is an interesting observation to set beside that of the sortie-stimulated fighter pilot: 'It was exactly like the movies!' Virilio's thesis is that war and cinema are inextricably linked – the fact that both involve shooting is more, it seems, than a weak metaphor.

By the time of the First World War, conventional methods of monitoring the progress of battle were proving inadequate. Static maps could not keep up with the movement of mechanized armour, and direct observation became impossible as the range of artillery and radio communications widened the field of operations. Instead, new systems had to be deployed; guns were fitted with telescopic sights, aeroplanes with cameras, before they were fitted with bombs, according to Virilio. 'In industrialized warfare, where the representation of events outstripped the presentation of facts', he wrote, 'the image was starting to gain sway over the object, time over space.'[11] The picture overtook what it depicted, images of war *became* the war. The 'authentic' war was no more to be found on the field of conflict than the 'authentic' version of a film is to be found on its studio set. Virilio quoted the account of Colonel Jack Broughton's *Thud Ridge*, an account of the Vietnam War as seen from the point of view of the pilot of an F-105 Thunderchief: 'Broughton writes: "I have bombed, and seen my troops bomb, on specific targets where I have watched the bombs pour in and seen the target blow up, with walls or structures flying across the area, only to be fragged right back into the same place because the film didn't look like that to the lieutenant who read it way back up the line." People used to die for a coat of arms, an image on a pennant or flag; now they died to improve the sharpness of a film. War has finally become the third dimension of cinema.'[12] Weapons have even become a form of photography, argued Virilio, with the heat of the Hiroshima atomic bomb imprinting pictures of its victims on walls. Particle beam weapons are like cameras in reverse, emitting beams of light to dematerialize their subjects. Members of the aircraft carrier *Nimitz* reportedly told *Libération*: 'Our work is totally unreal. Every now and then, fiction and reality should get together and prove once and for all that we are really here.'[13] Just as the Gulf War fighter pilot said, war is just like the movies.

'His prose tracking almost like a film camera,' wrote the military historian and Sandhurst lecturer Richard Holmes, 'Clausewitz describes the sensations a novice might experience as he moved across the battlefield.'[14] Far from trying to vindicate Virilio's position (whose work he did not cite), Holmes was trying to show how 'empty and drab' the battlefield was compared with the 'colour and texture' given to it by the 'rich palette of artists, writers and filmmakers'. In Clausewitz's words, 'the light of reason is refracted in a manner quite different from that which is normal in academic speculations'. Nevertheless, the 'real' battlefield does not turn out to be any more real than any other aspect of the war's conduct. Holmes quoted soldiers who described the battlefield as 'empty'; once a soldier is within enemy range he 'sees nothing, there is nothing to be seen'. 'People in England knew more of [the war's] progress from day to day than we did,' wrote a German officer in 1918. On the one hand, war confronts the ultimate reality, the binary fact of death. On the other, the closer to the battlefield the soldiers get, the less real it seems to be; an object visible from afar seems to vanish as it is approached.

And so, in the Gulf War, the battlefield was even less visible. It could apparently be most clearly seen through a night-sight or on a computer screen. The direct experience of the event itself did not seem to make sense. 'In a ground war, sound and vision no longer marry up,' wrote Robert Fisk in a powerful piece published by the *Independent on Sunday* on the day after the Allied ground attack commenced. 'To a soldier in his first ground battle, reality is broken apart like a film whose soundtrack has wobbled out of sync. Shells burst in silence; explosions have no source. A fighter-bomber will attack a distant target, bathe the terrain in fire and twist away into the sky without the slightest sound. Only later, as the smoke drifts away, are the senses reconstituted with the appropriate decibels.'[15]

* * *

On a cold, sunny morning in December 1988 I was sitting in the office where I am now writing these words. Absorbed by my work, I gradually became aware of a growing crescendo of police and ambulance sirens. Being louder and more insistent than usual, I walked out into the street to see what was happening. At one end, blocking a road bridge over the railway line leading into central London, I saw the cause, a jam of emergency service vehicles: police cars, ambulances, fire engines plus, just arriving, large vans with legends such as 'Mobile Accident Control Centre' written across their sides. I walked up to see what was happening. There seemed to be some commotion on the railway line. Walking onto the bridge, I could see that a train had derailed, and small groups of people were being helped out of the carriages and being escorted up a steep embankment. It was, by that time, quite quiet. Everyone seemed to be going about their business in an orderly way.

What I had witnessed was the immediate aftermath of the Clapham rail crash, the worst railway disaster in Britain for 20 years. Thirty-four people were killed and hundreds injured. But only on returning to my house and watching the first news bulletins that began to appear later that morning did I even realize it was a 'disaster'. The scene as witnessed it from that bridge had little meaning. By the evening, order had been established. Camera crews were taking shots from a police cherry picker crane, officials patrolled wearing jackets identifying their role, the trees and telegraph poles were draped with cables. At the centre of all this, immersed in a pool of white arc light, isolated from the world by police cordons, lay the wreckage, now a beautifully-lit tableau, soaked in the spraying sparks of cutting equipment, being gradually dismembered by powerful lifting gear. It had become a set.

What I had witnessed that day 'for real' at the end of the road was something quite different from what I had seen on television – and it was on television that, ultimately, the event happened, for me as much as for anyone else. This does not mean that everyone had the same experience. Nor does it imply that the experience of the victims or rescuers was any the less authentic and, for that matter, tragic or heroic. It simply means that the Clapham railway disaster and the collision of three trains on the stretch of line leading into Clapham Junction on the morning of the 12 December, 1988, were not at one level the same events. The difference between the two was economically expressed by Charles Jencks in another context. When asked about the Greek colonels' coup, which had taken place while he was visiting Athens, he replied: 'I didn't know anything because I had only *been there*'[16] An event such as that, like a major transport disaster, simply does not exist solely in the physical phenomena associated with it.

For Jean Baudrillard, the War That Did Not Take Place took place not on the deserts of Arabia; it took place on screens of the world's TV sets. 'Our strategic site,' he wrote in the article published just before the outbreak of hostilities, 'is the television screen, from which we are daily bombarded.'[17] It was a war that was constituted in its media coverage. It was a 'live' war, taking place before our very eyes as the global village network beamed the action direct into our homes. 'It has been the greatest bombardment in the history of television,' wrote Noel Malcolm in the *Daily Telegraph*, using what had become a familiar metaphor. 'Wave after wave of it has poured into our homes, and no element of the modern arsenal has been spared.'[18] The TV audience seemed to have the same view of the battle as the participants, watching the same videos of smart missiles seeking out their targets, watching the same night-sight pictures of tanks being burnt out by the heat of exploding tank turrets, watching the ungainly trajectory of Tomahawks taking off from American warships for invisible strikes, watching, watching, just as the commanders were themselves just watching. We did not all see the same thing, but it was the same thing being seen – images, representations, pictures, never the 'thing itself'.

Baudrillard's argument, echoed throughout the radical French press, was that there was no 'thing itself', no 'real' war depicted in those images. It was a postmodern war, a war where there is no reality, just, in Baudrillard's language, a simulation of it. 'The "live war" is a fantasy,' wrote Serge Daney in *Libération*. 'What is "live" is the *mise en scène* of all the information, true, false and omitted.' The phrase *mise en scène* is a popular one among French intellectuals and those they influence. It is hard to translate. It has its origins in theatre, where it means the arrangement of the stage, but it was picked up by cinema critics, who gave it a more sophisticated meaning. For them, the *mise en scène* can be thought of as the mechanism of cinematic illusion, the relationship between what the audience sees and the staging and framing that produces it. Baudrillard's argument is that the Gulf War was a staged event in which the stage was not to be found in the theatre of war. It was, he wrote, a 'virtual' war, breaking the classical chain of logic from virtual to actual by moving in the opposite direction. It was a war constantly examining itself in the 'mirror' of the television screen: mirror, mirror on the wall, was I the fairest war of all? It was a war never declared, never truly joined, never properly finished. It was what we had wanted it to be, never what it was. 'In our fear of the real, of anything that is too real, we have created a gigantic simulator.'[19]

Baudrillard's argument has its origins in a series of essays he has written that aim to confront what he sees as the 'obscenity' of postmodern life, of the chaotic ruins of the modern that lay in the rubble of the Pruitt-Igoe. Whereas critics like Charles Jencks are upbeat about the postmodern experience, celebrating its pluralism and playfulness, Baudrillard is disgusted by it. At the centre of his disgust lies the discovery that reality no longer exists – that, quite the reverse, it has become a fiction. The concept of an independent, objective reality was an assumption of the modern age that could not be sustained in the face of the technological and economic developments of the late twentieth century. Even critical theory and structuralism had operated on the principle that, at the end of the day and the conclusion of the theory lay the ultimate vindication of what was being argued, the basis of truth: reality. No more. According to Baudrillard, reality, like the Gulf War, is a simulacrum, a perfect copy that has no original.

Baudrillard began his classic essay on postmodernity, 'Simulacra and simulations' with a quote from Ecclesiastes: 'The simulacrum is never that which conceals the truth – it is the truth which conceals that there is none. The simulacrum is true.'[20] The 'finest allegory' of simulation for Baudrillard was – as it should be for any authentic analysis of postmodernism – an idea from a Borges short story about imperial cartographers who drew a map of their empire that was so accurate it exactly covered the territory it mapped. 'Abstraction today is no longer that of the map, the double, the mirror or the concept', Baudrillard commented. 'Simulation is no longer that of a territory, a referential being or a substance. It is the generation by models of a real without origin or reality: a hyperreality.' Whereas in the Borges story the map decayed along with the empire it depicted, the territory it covered eventually reasserting itself, it is the territory that is now disappearing, the map that survives. Worse, the empire is now capable of constructing the real so that it fits in with the map.

Like all good Marxist critics (though Baudrillard abandoned Marxism, which must remain a modernist condition, an optimistic state which entails that, underneath all the simulation, the reality must somewhere survive), Baudrillard saw the breakdown of reality as developing out of a series of distinct historical stages. These stages concern the history of signification, the way signs are used – the way, say, a picture relates to what it depicts, or a sentence refers to whatever it is about – and they plot the gradual separation of the sign from what it signifies, the separation of culture from nature, 'truth' from reality. In the first stage, the sign reflects a basic reality, in the second, it hides that reality, and in the third it hides the *absence* of that reality. These stages are not all that easy to plot against particular historical changes, and Baudrillard does not make much effort to explain them. Their importance lies in providing a run-up for the final, and for Baudrillard terminal, stage, the stage of pure simulation, when signs cease to signify anything real. Culture is about producing signs that signify nothing, that have only spurious significance. In the consumer society, industry is capable of producing just about anything anyone could want: machines that clean dishes, lamps that produce a skin tan, beer that does not make its drinker drunk, and fatless fat. Technology has enabled basic needs to be more than adequately satisfied for the majority of the population, which leaves desire as the only motivation to consume. The good thing about desire as an alternative to need is that it is the motivational equivalent of a blank cheque: manufacturers can fill it in with whatever they want. It is the role of advertising to generate desire, which it does, not by promoting the usefulness of what it is trying to sell, but by manipulating its meaning. Because it contains coconut, a chocolate bar is sold by associating it with tropical paradise, itself something with no geographical existence, but meaningful only with respect to Hollywood films or holiday brochures, themselves linked to late nineteenth-century novels and colonial exploits and so on and so on. The chocolate bar itself becomes the signifier of this web of associations; that is what it 'means'. This is not to say that its advertisement is in some sense misleading or false – without that 'meaning', there would be no product, the chocolate bar would be meaningless. Desire, then, becomes a matter of meaning, of the manipulation of signs, and what sustains the consumer society is the ability of manufacturers to perform this manipulation, to manufacture meaning. The same applies to culture as a whole, not just the manufacturing of consumer goods – politics, entertainment, the arts, everything that is the product of human effort. The result is Baudrillard's hyperreality: 'When the real is no longer what it used to be, nostalgia assumes its full meaning. There is a proliferation of myths of origin and signs of reality; of second-hand truth, objectivity and authenticity. . . . And there is a panic-stricken production of the real and the referential.'[21]

* * *

In the spring of 1992, a Magic Kingdom was opened in the heart of Europe's first modern republic. On a 5,000-acre site on the outskirts of Paris, obligingly provided by the French government at 1971 agricultural prices, Disney opened its European theme park. Given the French enthusiasm for protecting itself from alien cultural influences with institutions like the Academie, it was puzzling that a visitor with cultural imperialism (or, more accurately, cultural exploitation) so obviously on its mind should have been allowed in without so much as a baggage search. Nevertheless, President Mitterrand's Socialist government, which even appointed a Disney minister to help coordinate the project, seemed impressed by commitments to model Cinderella's Castle on drawings in *Les Tres Riches Heures du Duc de Berry* and to include an attraction based on Jules Verne's *20,000 Leagues Under the Sea* – neither of which entailed much compromise, since there is a 20,000 Leagues Under the Sea ride at the Florida Disney World and Cinderella's Castle ended up looking like all the other Cinderella's Castles.

In the end, all of the EuroDisney Resort was exactly like the other theme parks – indeed, to enter one is to enter them all, to walk up the same Main Street USA, explore the same Adventureland, pioneer the same Frontierland. There are attractions unique to each park, particularly at Florida's Disney World, by far the biggest, nearly twice the area of Manhattan, with two theme parks in addition to the Magic Kingdom: EPCOT and MGM/Disney Studios. But it is the Magic Kingdom that is the authentic land of make believe, surrounded by an embankment to 'seal the magic in', as Disney employees say (not, apparently, to keep the gate crashers out). It is a sovereign state entered by a 'Disney Passport', that even has its own currency, the 'Disney Dollar'. Both passport and currency are honoured in any Disney theme park in the world.

And what a world: beautifully tended lawns decorated by magical topiary, choruses of 'When you wish upon a star' floating out from the flower beds; you behold in awe the cleanliness of it all, a world smoother, more perfectly groomed (yes, even perfection must now yield to degrees) than a putting green. You may be queuing in 90 degrees of humid Floridan heat or centimetres of French drizzle, forced into line by an adapted cattle crush or stuck in a push chair pile-up at the Space Mountain intersection, but nobody cares. The pavement is clean enough for open heart surgery and birds still twitter in the trees. It is the totalitarian state that worked: no class, no poverty, no perplexing free market (Coca Cola products are the only permitted soft drinks), orderly queues for food, no sight of security personnel, and no books. Truly, no one here is being forced to be free, though they do have to pay.

This is the world that Baudrillard feels is the 'perfect model of all the entangled orders of simulation'.[22] 'Disneyland is there to conceal the fact that it is the "real" country, all of "real" America, which *is* Disneyland. . . . Disneyland is presented as imaginary in order to make us believe that the rest is real, when in fact all of Los Angeles and the America surrounding it are no longer real, but of the order the hyperreal and of simulation.' Baudrillard's argument is that California's Disneyland, like all theme parks, is there as a substitute for a diminishing sense of reality – indeed, to reinforce the sense that there really is an embankment surrounding the imaginary and preventing it from bleeding out into reality – a reassurance that 300 million of us have sought at the Florida Disney World alone.

Umberto Eco, the Italian semiologist and author of *The Name of the Rose*, also visited Disneyland, and it seems to have left him with very much the same impression. 'Disneyland,' he wrote in his collection of essays *Travels in Hyperreality*, 'tells us that technology can give us more reality than nature can.'[23] He travelled the wild river in Adventureland, watching the alligators bask on the banks. Then he travelled on the Mississippi, one of the originals of which Adventureland was a copy, only to be disappointed not to see the promised alligators basking on the banks – 'you risk feeling homesick for Disneyland, where the wild animals don't have to be coaxed'.

Just as the industrial revolution produced a mass migration from farm to factory, the postindustrial revolution seems to be producing a mass migration from factory to fantasy. Just as dirt tracks at the EuroDisney Resort construction site were resurfaced with simulated dirt track, so maintenance-intensive Victorian houses are done up in order to be conserved. Just as Disney has turned imaginary fairy castles into buildings, so tourism (which some argue will be the world's biggest industry by the turn of the century[24]) has turned Europe's buildings into imaginary fairy castles.

Paris is itself a theme park. Not only do its drivers obligingly supply the white knuckle rides, but it has become the venue for possibly the most spectacular architectural display since the war, a show more extravagant even than Disney's, designed to reinvent the modernist enterprise, but succeeding only to demonstrate its demise. As part of the celebration of the bicentenary of the French Revolution, President Mitterrand instituted a series of *grands projets* designed to create a set of monuments celebrating France's achievements: a great glass pyramid covering the entrance to the Louvre; a soaring square arch at La Défense, extending the axis of the Champs-Elysées; a people's opera house at the Bastille; an urban park at la Villette. All of these were rapturously received by architectural critics as examples of the sort of vision that postmodernism had destroyed with its petty, plagiaristic, pandering aesthetic.

Perhaps the scheme that best represented the spirit of the *grands projets* was the Parc de la Villette. This was a park intended to break all the rules. There were to be no flowers or fountains, but gardens of artificial fog; no bandstands, but rather buildings purposefully designed to have no purpose. Once the site of the capital's slaughter houses, the only cows to be slaughtered in the Parc were to be the sacred sort, where the conventions of urban living were to be taken apart in an attempt to create the model of the twenty-first-century city space. The architect, hand-picked by President Mitterrand, was Bernard Tschumi, a so-called 'deconstructivist' architect who set about creating buildings that would have no single meaning, because, as all deconstructivists have discovered, there is no single meaning to any text. So he scattered the park with little red metal box buildings he called *folies*, not because they were like the sorts of follies that might be found in an English park, but because they were mad – architecturally mad so that visitors could better appreciate the constructed nature of normality. They were buildings built to service no particular function, because their function would come from the way visitors to the park interacted with them.

The result was not just a brilliant intellectual exercise – one sanctified by a written testimony from the god of deconstruction, Jacques Derrida – but a very popular city park. But, like the other *grands projets*, it was also a self-defeating simulation, a piece of expensive hyperreal estate. It was stylistic exercise, a play with the aesthetics, the look, of high modernism – futuristic, mechanistic, pure, uncompromising. All the *projets* were supposed to have some social purpose, except not one that was recognizably to do with serving what in the modern era would have been seen as obvious material needs: a people's opera house, a people's park, a people's art gallery entrance, a people's international conference centre, a people's library, a people's finance ministry. No people's housing or hospitals. Mitterrand's modernism, his vision, his monumentalism resulted in an architectural toy town that, from a position of pessimistic postmodernism, was as fabulous, as nostalgic and ultimately as fake as Cinderella's Castle. It seemed to support Baudrillard's gloomy intuition that all anyone can create is a fake; everything is stuck in the endless play of weightless signs.

Between San Francisco and Los Angeles, Umberto Eco was able to visit seven wax versions of Leonardo da Vinci's *Last Supper*. 'The voice has warned you that the original fresco is by now ruined, almost invisible, unable to give you the emotion you have received from the three-dimensional wax, which is more real and there is more of it.'[25] The 'original' of da Vinci's work was painted on the wall of the Monastery of Santa Maria delle Grazie in Milan using an experimental technique which, unfortunately, has led to the painting's gradual and irreversible disintegration. Set at the end of the Monastery's Refectory, the lines of the wall are projected into the space of the painting to create what Ernst Gombrich described as an 'extraordinary illusion of reality'. 'Never before had the sacred episode appeared so close and so lifelike. It was as if another hall had been added to theirs, in which the Last Supper has assumed tangible form.'[26] Except the simulated hall is slowly disappearing.

The *Last Supper* is one of the most widely reproduced images in the world. Yet most of the copies of it are inaccurate, because they show not the fresco on the refectory wall, but an untarnished version, its colours more vivid than they ever could have been on the refectory wall.[27] Since the original is such an inadequate representation of Leonardo's great picture, why bother preserving its faint remains – why not, indeed, paint it over with a sparkling new version, using, if necessary, the techniques and pigments Leonardo would have used? It would at least be more like the Last Supper depicted in all the reproductions – a more accurate original of the copies.

In 1990, Polaroid announced the introduction of a new process for creating what it called 'museum quality reproductions'. 'Using a combination of very-large-format instant photography and state-of-the-art digital image processing', boasted the promotional literature, 'the Polaroid process has achieved a level of colour accuracy and faithfulness to detail never before available.'[28] The results were, undoubtedly, startlingly 'real'. Even though the process was unable to simulate the texture of the paint, the brilliance of the colour, the clarity of the detail and the sheer size and the presentation in 'hand-crafted solid wood' frames result in convincing replicas. 'Imagine having the opportunity to own and actually live with works of proven importance by artists like Monet, Cezanne, Sargent, Homer and more. Each day will bring new discoveries in nuances of colour and technique.'

For postmodernist critics, the *Last Supper* reproductions and Polaroid replicas are symptomatic of mounting fear of losing the reality that sustained the modern world in the absence of religious authority. Auction house prices, tourism, theme parks and even wars are ways we have developed for fooling ourselves that such a reality still exists. We are like cartoon characters who have walked off a cliff edge and, still suspended in the air, have suddenly realized there is nothing beneath us. This is the 'crisis' of postmodernity, and it is technology that has produced it.

* * *

John von Neumann's 'universal constructor' was, as we saw earlier, a conceptual machine designed to explore the idea of self-reproduction. When he first thought of the idea, he imagined what he called a *kinematic* model of a robot swimming through a lake filled with components out of which it constructed a copy of itself. He imagined artificial arms for manoeuvring these components, devices that would fuse them together and cut them apart, sensing elements to locate the components, girders that would provide a framework and so on.[29] Von Neumann never imagined that such a machine would be built, but in the most advanced Japanese factories something like it is close to being realized: what might be called the universal factory, a machine capable of making any product. And what has made this possible is the introduction of that other universal machine, the computer, into the production process.

One can too easily be overawed by the Japanese industrial machine into thinking that it is both perfectly efficient and technologically advanced. In fact, much of the country's economic success rests on its small subcontractors, the thousands of tiny companies that make the basic components assembled by the manufacturing giants. On a visit to Tokyo, staying in the Osaki district that forms one of the centres strung along the city's Yamanote railway loop, I strolled up a small alleyway behind the modern 'executive' hotel to admire the huge Sony factories that overlooked the area. They were not particularly impressive buildings, just blank-walled boxes with the familiar logo stuck on every visible surface. The only distinctive feature was the basketball court each had on its roof. As I walked round one of the factories, I encountered a collection of wood and corrugated iron huts, ramshackle as rooks' nests, leaning against a perimeter wall. I peered through one of the doors into the gloom. It was like looking through a window onto another age. A group of faces were staring with rapt concentration into the flames of a forge, their faces illuminated by its glow. Around them could just be discerned the dull black forms of iron castings. It was a sight straight out of the industrial revolution, one that would have been commonplace in the Black Country of England in the mid nineteenth century. And just above this glowed the mouthwash blue fluorescent sign of Sony.

The mix of these tiny workshops and the global multinationals, the *zaibatsus*, has created a intricate and, in Western terms, almost antiquated industrial structure. Nevertheless, it is this structure that has formed the foundation for Japan's ambition to build the world's first true information society. In 1980, Japan's Industrial Structure Council reported that the country's 'informization', its transformation into a postindustrial, information economy, would happen in two phases. The first phase, produced by the computerization of factories, offices and distribution systems, was already completed by the end of the 1970s. The second stage, the 'high-level information society', was scheduled to begin in the 1980s, and would mark the point when all the separate computerized systems would be brought together, creating a new, more flexible relationship between demand and supply, replacing mass production and standardized goods with bespoke production and personalized goods.[30] It was essentially the vision promoted by Media Lab and the computer revolutionaries, only applied to the production of physical objects, the manipulation of the material world.

The universal factory is central to the implementation of stage two. It will be like a three-dimensional computer printer, capable of fabricating just about any physical object from a computer model. Each product produced by this process will be a literal simulacrum, at least in Baudrillard's sense, a copy that has no original – or rather, only a virtual original, a computer simulation of the product, tested in its own virtual environment, which is downloaded to the factory floor for assembly by computer-controlled robots. The universal factory will be linked, via a computer network, to the universal shop, which will provide customers with suitably 'user friendly' computer aided design tools so that they can take pro forma products and tailor them to their individual needs and tastes, the information being sent to the factory to be fabricated for delivery to their doorstep.

Of course, it will not be quite that easy – the universal factory and shop are still just a 'computopian' dream. But the first manifestations of the underlying idea have already begun to shape the Japanese consumer market.

By the mid 1980s, its exports had so saturated world markets that Japan began to look for further room for expansion at home. The problem was that the domestic market was saturated, too; everyone owned at least one of everything – what they did not own, such as a second car or reasonably sized house, they could not own, because there was simply not enough space. One solution came from the designer Naoki Sakai, who was employed by multinationals like Nissan and Olympus to design beautifully detailed 'retro' products. If the past is a foreign country, then Sakai's designs were its tourist souvenirs, instant antiques for a nation which preserves nothing. For Nissan, he designed a noddy car called the Pau, for Olympus a fake Bakelite camera called the Ecru. Both were obviously contemporary (the Ecru had in-built flash and motor wind), but nevertheless were inspired by the technology of some pre-industrial, pioneering age when technology was not packaged in anonymous black boxes and sold throughout the world. There was, of course, no such age – at least, not until Sakai created the artefacts that recalled it. They were like the copies of Leonardo's *Last Supper*, authenticated by an original that did not exist.

Traditional manufacturing techniques would never have been able to cope with Sakai's designs. They were the first generation of universal factory products – limited runs, one-offs which did not rely on some universal notion of utility or function for their value. Their purpose was to mark the arrival of the postindustrial consumer society, the sort of society where the camera or car is no longer a means of recording images or of moving from one place to another, where the purpose of a product is explicitly bound up with what it signifies. We are already familiar with the idea of the car being a status symbol. Now even status is no longer a single, stable, objective condition that a product can symbolize. A poor person hires an expensive car to acquire the status – and then, perhaps, sets about finding the job or social reputation that would justify its ownership. A rich person continues to drive around in a tattered banger not to disguise his or her wealth, but to advertise that they need not advertise it. The whole process becomes an impossibly complex set of moves in the game of signs and symbols.

Following the success of the cars it had commissioned Sakai to design, Nissan decided to use its own design team to produce what was to become its most successful retro product, the Figaro. It only made 8,000 at a time, thus creating an artificial shortage. To balance demand with supply, it did not use the conventional capitalist practice of raising the price, but instead sold them at a relatively low fixed price by public lottery. The result was a car instantly transformed into a classic, valued for its rarity and for its celebration of design values validated by tradition and nostalgia rather than modernity and utility. Next year, of course, customers may be clamouring for a new type of tack-technology, celebrated for its disposability and primitivism; perhaps the following year, they will want organic design and natural materials, it does not really matter – what matters is that manufacturing is flexible enough to sustain the free play of signs. 'Our entire contemporary social system has little by little begun to lose its capacity to retain its own past,' observed Frederic Jameson, despondently, 'has begun to live in a perpetual present and in a perpetual change that obliterates traditions of the kind which all earlier social formations have had in one way or another to preserve.'[31]

* * *

The cover of the February, 1982 issue of the *National Geographic* magazine featured a photograph of the pyramids at Giza on its cover. There was nothing remarkable about the picture, it was a conventional treatment of a standard *National Geographic* subject; exotic, beautifully composed, richly coloured. However, it later emerged that the magazine's picture editors had manipulated the image using a computer. They had 'moved' the pyramids closer together, so that the horizontal picture would fit into the vertical space of the cover. As far as the magazine's editors were concerned, changing the photograph's composition in this way was no different to what a photographer might do by taking the picture from a different angle. However, critics were alarmed that the result was a deception – it was no longer what it purported to be, a photograph of the pyramids. Reality had been sacrificed in the cause of convenience.

The assertion that 'the camera never lies' is, of course, untrue. Photography may be, as one of its inventors, William Fox Talbot, put it, the 'pencil of nature', but the nature it draws has always been manipulated for the benefit of the drawing. Indeed, the earliest photographs were specifically composed with sets and studios to look like paintings. Even photojournalists admit that the composition and framing of their pictures can change the meaning – their works are, in other words, as 'authored' as the written reports that usually accompany them. Their 'truth' is a factor not of the process, but of the integrity of the photographer who took them.

Nevertheless, photography is accepted as being objective in a way that drawing is not. Being a mechanical process, there is a fixed relationship between the picture and what it depicts that is at the very least difficult to loosen invisibly, which is why totalitarian states painstakingly doctored photographs to erase dissidents from official history – it was not sufficient for Stalin to destroy all the pictures of Lenin with Trotsky, to ensure the latter's deletion; Trotsky had to be removed from the pictures to *prove* that he had not been there. 'The photograph's essence is to ratify what it represents,' wrote Roland Barthes. 'One day I received from a photographer a picture of myself which I could not remember being taken, for all my efforts; I inspected the tie, the sweater, to discover in what circumstances I had worn them; to no avail. And yet, *because it was a photograph* I could not deny that I been *there* (even though I did not know *where*).'[32]

With computers, there is no such ratification. The 'pencil of nature' has fallen into meddlesome human hands. Images, moving and still, can be manipulated without the manipulation being detectable, offering photorealism without the inconvenience of reality. Advertisers have used this not to produce misleading pictures but rather to show that reality itself is something they can change for the benefit of their customers. One famous example was a poster produced by the agency WMGO for the Virgin Atlantic airline showing the Hollywood sign changed to read 'Jollywood'. The manipulation was undetectable, yet the resulting picture would fool no one – indeed, the poster only worked because the image was such a familiar one. It was a joke, but nevertheless one that served to suggest Virgin's ability to create a reality to suit its customers.

'Photorealism' remains one of the goals of the computer graphics industry, and one they have already reached using the most powerful computers. The lure of photorealism is the simulation of reality, creating an image with all its attributes intact except one: its fixed relationship to what it depicts. It is the achievement of this level of simulation that shows how reality has itself become part of the artificial realm. The emergence of reality as a technological goal shows how the more it slips from our grasp, the more desperately we cling to it. 'Today, *reality itself is hyperrealistic.*' wrote Baudrillard. 'Now the whole of everyday political, social, historical, economic reality is incorporated into the simulative dimension of hyperrealism; we already live out the "aesthetic" hallucination of reality. The old saying, "reality is stranger than fiction" . . . has been surpassed. There is no longer a fiction that life can confront, even in order to surpass it; reality has passed over into a play of reality.'[33]

Notes

1 Published in English as 'The reality gulf', *Guardian*, 11 January 1991, p. 25.
2 *La guerre du Golfe a-t-elle vraiment lieu?*
3 Jean Baudrillard, *La Guerre du Golfe n'a pas eu Lieu*, Paris: Galilée, 1991.
4 *Sun*, 18 January 1991, p. 1.
5 *Observer*, 20 January 1991, p. 16.
6 *Independent on Sunday*, 20 January 1991, p. 10.

7 Frank Barnaby, *The Automated Battlefield: new technology in modern warfare*, Oxford: Oxford University Press, 1987, p. 1.
8 Barnaby, 1987, p. 107.
9 Glen Johnson, 'Simnet protocol opens the way to interactive battle simulation', *Digital Review*, 11 December 1989, 6 (49) p. 51.
10 Paul Virilio, *War and Cinema: the logistics of perception*, trans. Patrick Camiller, London: Verso, 1989, p. 15.
11 Virilio, 1989, p. 1.
12 Virilio, 1989, p. 85.
13 Virilio, 1989, p. 66.
14 Richard Holmes, *Firing Line*, London: Penguin, 1987, p. 148.
15 Robert Fisk, 'The bad taste of dying men', *Independent on Sunday*, 24 February 1991, p. 19.
16 In Martin Pawley, 'Post coffee table', *Blueprint*, 82, November 1991, p. 18.
17 Baudrillard, *Guardian*, 11 January 1991, p. 25.
18 Noel Malcolm, 'No hiding place from the war of words and pictures', *Daily Telegraph*, 24 January 1991, p. 12.
19 Baudrillard, *Guardian*, 11 January 1991, p. 25.
20 Jean Baudrillard, 'Simulacrum and simulations', in Mark Poster (ed.) *Selected Writings*, Cambridge: Polity Press, 1989, p. 166.
21 Baudrillard, 1989, p. 171.
22 Baudrillard, 1989, p. 171.
23 Umberto Eco, *Travels in Hyperreality*, London: Picador, 1987, p. 44.
24 For example, see 'Wish you weren't here', *Nature* BBC2 television, 25 November 1991, 8 pm.
25 Eco, 1987, p. 18.
26 E. H. Gombrich, *The Story of Art*, 15th edition, Oxford: Phaidon, 1989, p. 224.
27 The *Last Supper* example was developed in 'The shock of the neo', *Signals* produced by Illuminations Television for Channel 4, 1 February 1989.
28 'The Polaroid museum replica collection', Cambridge, Mass.: Polaroid Corporation, 1990.
29 William Poundstone, *The Recursive Universe*, Oxford: Oxford University Press, 1985.
30 Tessa Morris-Suzuki, *Beyond Computopia: information, automation and democracy in Japan*, London: Kegan Paul, 1988, p. 11.
31 Frederic Jameson, 'Postmodernism and consumer society', in E. Ann Kaplan (ed.) *Postmodernism and its Discontents*, London: Verso, 1988, p. 28.
32 Roland Barthes, *Camera Lucida*, trans. Richard Howard, New York: Hill & Wang, 1981, p. 85.
33 Baudrillard, 1989, p. 147.

11
REALITY

'We stand at the dawn of a new era,' wrote John Naisbitt and Patricia Aburdene in their best-selling *Megatrends 2000*. 'Before us is the most important decade in the history of civilization, a period of stunning technological innovation, unprecedented economic opportunity, surprising political reform, and great cultural rebirth.'[1] We are approaching one of the most dramatic cultural transformations in human history, argued Fritjof Capra, provoking 'not just a crisis of individuals, governments, or social institutions; it is a transition of planetary dimensions. As individuals, as a society, as a civilization, and as a planetary ecosystem, we are reaching a turning point'.[2] Meanwhile George Gilder has announced that the 'central event of the twentieth century is the overthrow of matter. In technology, economics, and the politics of nations, wealth in the form of physical resources is steadily declining in value and significance. The powers of mind are everywhere ascendant over the brute force of things.'[3] And it is a new form of physics, argued Paul Davies and John Gribbin, the science that gave rise to materialism, that is responsible for materialism's demise. 'During this century the new physics has blown apart the central tenets of materialist doctrine in a sequence of stunning developments.'[4]

If this is the beginning of a new era, the dawn chorus has become deafening. Books seem to be published weekly announcing the coming of the New Age, the 'new physics', the 'new science', the greatest shift in our conception of reality since, depending on your point of view, Einstein published the theory of relativity at the beginning of this century, since the birth of modernism in the middle of the last, since the founding of humanism with the French

Revolution at the end of the eighteenth century, since the beginning of the modern era in the Renaissance, since the beginning of human civilization or, if you are Timothy Leary, since the beginning of human evolution. Everyone senses change. The horses are whinnying in their stables, the dogs are restive, and the soothsayers are proclaiming the revelation of new sooths: we are at the cusp of the most momentous changes ever witnessed, a new era of human civilization, a shift in paradigms as dramatic as a reconfiguration of the stars.

You could put it down to a revival of that wild fin de siècle feeling that overtook the world at the end of the nineteenth century, amplified by the fact that this time we are approaching not simply the end of a century but the end of the millennium. But the accident of the decimal numbering system and traditions of Christianity hardly seem sufficient to bring about a 'turning point' of 'planetary dimensions'. It may explain the sales figures of books that announce such a turning point, but not the turning point itself. For that, one has to look at a shift that has been underway for much longer than the publishing phenomenon that has announced it: a shift in the foundations of science.

* * *

What is reality? 'You ask – what is the meaning of truth, the meaning of reality? Here it is', wrote Viscount Herbert Samuel with aristocratic assurance. 'The whole body of scientific achievement – massive, world-wide – brings confirmation.'[5] The great claim of science is the revelation of reality. It is not just that it shows the content of reality, it has shown that a reality principle, to borrow a term from Sigmund Freud, works. According to this principle, the study of experience by means of objective procedures is profitable. This is why it stands in opposition to religion and faith. In Christianity, the discovery of God in a church, on the battlefield, in one's deathbed does not mean that others can achieve the same by copying the actions and behaviour that led to the revelation – though, of course, it may help. In science, the same procedures should result in the same discovery irrespective of who carries them out.

On 23 March 1989, two chemists at the University of Utah, Martin Fleischmann and Stanley Pons, held a press conference to announce that they had discovered cold fusion. It was a scientific advance of potentially revolutionary importance – sufficient to feature on news bulletins across the world. It meant they had found a way of creating cheap, abundant, clean energy, something that a world newly alerted to the greenhouse effect was only too eager to accept as a no-pain substitute for fossil fuels. Fleischmann and Pons were both respected scientists, but their method of announcing the discovery, hastened by rumours of a breakthrough at Brigham Young University in the nearby city of Provo, was cause of some concern. Various mechanisms exist for announcing scientific results, but most of them are slow, painstaking and inefficient at attracting the media's attention. Fleischmann and Pons, thinking that their discovery was too important to disclose via conventional channels, decided to announce their results at a public press conference. They made big claims but gave few details. Nevertheless, the apparatus they demonstrated – a pair of electrodes immersed in a test tube – was so simple, and the materials used – palladium for the electrodes and water enriched with a form of hydrogen – were so familiar that a number of other laboratories attempted to reproduce Fleischmann's and Pons's results simply on the basis of what they saw on the television coverage of the press conference. Meanwhile, the University of Utah set up a laboratory to develop cold fusion technology.

To begin with, signs were hopeful. Several scientists reported that they had found signs of cold fusion, 'despite the jeers of mainstream physicists and the general dismissal in the mass media', as one alternative science journal put it.[6] But within a few months laboratories began to announce their failure to produce anything like the results claimed by Fleischmann and Pons. 'Supporters of cold fusion will cite anecdotal confirmations but they fail to mention the many groups who have reported negative results', wrote Frank Close,[7] the senior principal scientist at the Rutherford Appleton Laboratory and author of a book on the cold fusion race.[8] 'The research groups for cold fusion tend to be small and tend to be relatively "amateur" compared with the full-time, large-scale teams of laboratory scientists who have, almost universally, seen nothing.' At the end of June, 1991, Utah University's Cold Fusion Institute was closed, and with it the dream of cold fusion, at least as imagined by Fleischmann and Pons.

The whole episode was instructive, because it showed the fine line science treads between objectivity and conformity. On the one hand, the scientific world shows an impressive commitment to standard procedures. Years of work can be destroyed by just a few experiments, and not even the reputation of a Fleischmann or Pons, nor wish-fulfilment, can prevent it happening. On the other hand, because work is so vulnerable, there is enormous pressure to conform to tried and tested theories. Very powerful orthodoxies emerge, and scientists question them at their peril.

Peter Duesberg, a molecular biologist at the University of California at Berkeley shocked researchers by announcing in 1987 that AIDS was not caused by the virus HIV. When he made the claim, the role of HIV was just about the only 'fact' that the scientific world had established about the disease since its discovery in 1981. He argued that no connection between HIV and the disease had been established, and that there was plenty of evidence that no connection existed (for example, he said that the presence of HIV antibodies in AIDS sufferers indicated that the body was able to fight the disease). He and his supporters saw themselves as being relegated to a 'dissident' group, denied funding and rejected by respectable journals. 'Duesberg appears to be revelling in some peculiar form of intellectual self-indulgence,' argued the *New Scientist* in an editorial. 'He is not nurturing scientific debate, but he is wasting the time of researchers.'[9] In September of 1991, two independent groups, one in Canada, the other in Britain, produced results that showed the connection between HIV and AIDS to be at the very least more complex than had been assumed.[10] Alarmingly for orthodox researchers, it revived Duesberg's ideas and also supported the protests of other dissident scientists who claimed they were being denied their fair share of drug company and government grants.

There are many controversial examples of non-conformism in modern science – Fred Hoyle's belief in the extraterrestrial origins of life, Jacques Benveniste's claim that solutions retain a 'memory' of molecules they once contained, Rupert Sheldrake's belief that memory is an inherent feature of nature. Such dissidents, all scientists, tend to regard themselves as oppressed minorities, the casualties of entrenched interests and sclerotic mind-sets. Their opponents, on the other hand, tend to regard the dissidents as mavericks and paranoiacs, sticking to a position out of sheer obstinate pride. Either way, dissidence demonstrates that science is not quite the precise, unambiguous business that non-scientists often imagine it to be. Results need to be interpreted, which makes quite basic questions, such as the cause of AIDS, hard to settle once and for all even when they have been settled with a reasonable degree of agreement. It is not always easy even for scientists to tell whether their theories are made valid because they accord with reality, or because they accord with other scientists and funding agencies. Lyotard's argument in the *Condition of Postmodernity* was that, with the collapse of the modern 'metanarrative' of reality, science is sustained by the 'performativity' of theories, their ability to generate more scientific work. 'In the discourse of today's financial backers of research', he wrote, 'the only credible goal is power. Scientists, technicians and instruments are purchased not to find truth, but to augment power'.[11]

However, fractured though the 'metanarrative' might be, the most remarkable feature of science even now must be its unity. Lyotard argues that it is splitting apart into separate disciplines with no means of communicating with each other. But thousands of quite independent organizations in countries often hostile to each other engaged in often secret work are all agreed that they are participating in what is effectively the same enterprise. They use roughly the same method based on roughly the same theoretical framework – Newtonian mechanics, Einsteinian relativity, quantum physics, Darwinian evolution and so on – and exchange their ideas through the same journals and at the same conferences.

It is this fact that seems to support Viscount Samuel's assertion that 'the whole body of scientific achievement – massive, world-wide' confirms the meaning of reality, even more so now than when he wrote it in 1958. Given its size and diversity, science is remarkably homogenous, remarkably consensual, remarkably stable. This does not mean it is as loftily detached from the fray of politics as it likes to imagine itself to be, nor that its international success is independent of its power to produce profits, nor that different cultures do not have different approaches to the conduct of science. The achievements of science reflect the preoccupations of the society that sponsors it – AIDS shows this clearly, with a large proportion of Duesberg's argument being based on his suspicion that HIV was isolated as the cause because it suited the drug companies to have a single, identifiable culprit for which they could develop a marketable vaccine. But still, for all its ruptures, dismemberments and disorders, it remains a single, whole body of knowledge that seems to have withstood the twentieth century better than most.

The 'alternative' science movement regards this unity as a product of science's exclusivity. It is, they argue, a closed club that selects only those ideas that will best sustain it. Why no room for astrology in cosmology, why exclude aromatherapy from medicine when it works for so many, why shun divination when there are so many reports of its effectiveness? Why are no scientific research funds spent on investigating ESP, out-of-body experiences, telekinesis? The simple answer is that if any of these are right, then all of science is wrong. Science only admits into its canon those theories that can be rejected by it.

In December of 1919, Sir Arthur Eddington of the Royal Astronomical Society organized expeditions to West Africa and Brazil. It was a trip straight out of the Victorian age of exploration, but with an important difference: the scientists were there to study the stars, not colonize the landscape. During a total eclipse of the sun, they observed how light from the stars that were revealed around its rim was apparently deflected by the sun's gravitational field. This provided the first confirmation of Einstein's theory of relativity, which had been published in 1915 – and at a stroke undermined Newton's theory of gravity, a theory that had persisted for over two centuries. 'This, I felt, was the true scientific attitude', wrote the philosopher Karl Popper later. 'It was utterly different from the dogmatic attitude which constantly claimed to find "verifications" for its favourite theories. Thus I arrived at the end of 1919 at the conclusion that the scientific attitude was the critical attitude, which did not look for verifications but for crucial tests; tests which could *refute* the theory tested, though they could never establish it.'[12] What impressed Popper was that the great edifice of Newtonian theory could be toppled by the ideas of a young, then relatively unknown physicist. Scientific hypotheses, such as the theory of relativity, are accepted as the product of human inspiration, but what sets them apart from other hypotheses is that they are *falsifiable*, any scientist can potentially prove them to be false simply by showing that they fail to match up to observations. Any theory that cannot be tested to destruction is not a theory. Science is not whatever scientists want it to be; it is forged against the hard anvil of observation. However, as Einstein was himself to be instrumental in discovering, this anvil is not as well tempered as physics had for centuries assumed it to be.

* * *

Quantum mechanics has been described as the most successful scientific theory ever, the foundation of modern physics and the basis of the technology of our times. Yet in 1935, Albert Einstein, whose work led directly to the theory's formulation, declared that it must be fundamentally flawed. If correct, it would devastate the most cherished assumptions of science. It would mean that there is one scientific law for the large, and another for the very small, that the universe science had assumed to be ultimately fixed and certain is really an uncertain figment of our imaginations. It would mean that reality is not the product of universal laws but of perception. 'An inner voice tells me that it is not yet the real thing', Einstein wrote to the physicist Max Born in 1926. 'The theory produces a good deal, but hardly brings us closer to the secret of the Old One. I am at all events convinced that *He* does not play dice.'

The first signs that God may be playing games with the classical scientific view of the universe came at the turn of the century. Since 1801, light was thought to take the form of a wave. All the experiments seemed to prove this. Light produces interference patterns as it passes through gaps in a barrier – something that would never happen if it was made up of particles. But at the end of the century, new instruments were revealing phenomena that seriously undermined this view. The German physicist Max Planck had established (while taking tea with a colleague, we are rather picturesquely informed by the physicist Werner Heisenberg[13]) that the mathematical formulae that best explained the absorption of energy by a 'black body', an experimental object that absorbs any radiation it is exposed to, entailed the existence of discrete 'quanta' of energy. Atoms, he assumed, had a sort of inner harmonic which meant that they could only absorb or emit energy in discrete amounts. In 1905, Einstein reinterpreted Planck's findings by applying them to another troublesome phenomenon called the 'photoelectric' effect, which concerned the rate at which light shone on a metal plate displaced electrons on the plate's surface. His conclusion was that light in particular and all forms of radiation in general must be made up not of waves but of quanta – particles which, in the case of light, he called 'photons'.

Worse was to come. In 1913, Niels Bohr showed that this ambiguity is not confined to light – that the stability of the atom could only be explained using this same quanta-based mathematics, a discovery which implied that electrons perform the famous 'quantum leap' trick, jumping between different orbits around the atom's central nucleus without at any time being anywhere in between. In the mid-1920s, the French nobleman Louis de Broglie had started to formulate the idea of 'matter-waves' on the assumption that if waves could be particles, then particles could be waves. In the process he introduced a perplexing duality in the very basis of matter and radiation that Erwin Schrödinger reinforced by successfully applying equations developed to describe conventional waves to these matter waves. The result was that a subatomic particle's state came to be described not by the usual formulae representing its position and speed but by a 'wave function', a formula that computed the probability of the particle being at any particular point at any particular time.

At around the same time, Bohr and others began to try to solve these equations not by treating the waves as real, physical phenomena but as waves of probability. It was a startling strategy, one implying that the electron – indeed, any subatomic particle – was not in one place at one time that the wave function helped determine, but actually smeared across a field of probability in a way that could only be described using statistics. 'It introduced something standing in the middle between the idea of an event and the actual event,' wrote Werner Heisenberg, 'a strange kind of physical reality just in the middle between possibility and reality.'[14] Furthermore, this strange kind of physical reality was fundamentally different to the sort encountered above the subatomic scale – and it took a completely different sort of mathematics to describe it.

In order to sort out the accumulating anomalies and peculiarities that came cascading out of the study of this newly discovered subatomic world, a group of physicists, including Schrödinger and Heisenberg, came to the Copenhagen Centre for Theoretical Physics at the invitation of Niels Bohr. The result of this consultation was not a theory so much as an interpretation of how the various developments could best be understood – one, however, that confounded both scientific dogma and common experience.

According to the Copenhagen interpretation, as it came to be called, a subatomic 'system' (in other words, the object of study, such as an atom) is fundamentally different from, but dependent upon, the system used to measure and observe it. The former works according to one set of physical laws, those of quantum mechanics, the other according to the conventional 'classical' laws set out by Newton and Einstein. However, only when the former is measured and observed by the latter can the former be said to exist as an actual phenomenon.

Bohr described the wave-like and particle-like characteristics of such a system as 'complementary', they are two aspects of the same thing that are impossible to behold simultaneously. According to Heisenberg's 'uncertainty' principle, the result of this complementarity is that there is only so much to be known simultaneously about both a subatomic particle's position and its momentum – to be more certain about one is to be less certain about the other, and to be certain about both is impossible. The important, and mysterious, point about the uncertainty principle is that it is not the result of imperfect measurement; there is simply nothing more to know, the uncertainty lies in the system itself, not just in the measuring of it.

In 1935, Schrödinger imagined an experiment that aimed to show what the Copenhagen interpretation meant to our conception of reality. The experiment would be conducted in a small gas chamber, sealed so that it was impossible to see inside. A piece of radioactive material (too small to be harmful) would be placed next to a Geiger counter, which, upon detecting radiation, would set off a trigger mechanism that would release the poison gas. And into this chamber, Schrödinger rather cruelly imagined placing a cat, which would be left there for sufficient time for the radioactive sample to have a 50/50 chance of emitting a radioactive particle. What would happen? The obvious answer is that the cat would either live or die, depending on whether or not the radioactive sample had decayed. But according to the Copenhagen interpretation, until the chamber is opened and the outcome observed, the cat would be both dead *and* alive, or, rather, in what quantum physicists call a 'superposition' of both states.

This startling conclusion is the result of the whole process being initiated by a quantum effect, the release of a particle, say an electron, by the radioactive material. According to the Copenhagen interpretation, while the box remains sealed, the 'real' state of the system is a superposition of its two possible states, mathematically described using the statistical wave function. At the moment the box is opened, this superposition collapses into one position. And only then would the cat be out of the bag. 'The transition from the "possible" to the "actual" takes place during the act of observation', wrote Heisenberg. 'If we want to describe what happens in [a subatomic] event, we have to realize that the word "happens" can apply only to the observation, not to the state of affairs between two observations . . . we may say that the transition from the "possible" to the "actual" takes place as soon as the interaction of the object with the measuring device, and thereby with the rest of the world.'[15]

Einstein could not accept such a conclusion. It undermined the unity of science, since it implied there were two quite separate domains, that of the subatomic realm and that of the observer. It denied the independent existence of the material world – the observer was no longer a passive onlooker but an active participant. It consigned actuality to a matter of probability, the dice game that Einstein so famously said that God would not play ("Stop telling God what to do", Bohr is said to have retorted[16]). Ultimately, it threatened to turn physics into a set of theories about observation rather than about the real nature of the universe. 'Does the moon only exist when you look at it?' he asked his colleague and biographer Abraham Pais.[17] There must, he argued, be more to it, some 'hidden variables' that explained such perplexing surface phenomena.

To prove his point, Einstein, along with his colleagues Boris Podolsky and Nathan Rosen at Princeton (where he had settled after fleeing Nazi Germany), published a paper in the American journal *Physical Review* in 1935 entitled 'Can quantum mechanical description of physical reality be considered complete?' The paper contained a 'paradox', called the EPR paradox after the initials of its authors, designed to show that the 'quantum

mechanical description' (in other words the Copenhagen interpretation) could not be complete.

Einstein took the case of a pair of subatomic particles that
bounce off each other. According to the law of conservation of
momentum (a classical law of physics that had survived even
quantum mechanics), the momentum of one particle could be
worked out by measuring the momentum of another, assuming,
as Einstein did, that neither had in the meantime encountered
any other obstacle. Suppose, then, that an observer decides to
measure the speed or position of one of these particles. That
particle's speed or position will thenceforward be predictable.
This would imply that the other particle's speed or position will
thenceforward be predictable. However, since the Copenhagen
interpretation insists that the observation of a subatomic particle's
speed or position in some way effects what can be known about
it, it must either be argued that nothing is known about the other,
unobserved particle's position or speed, or that the observation of
one particle can have some effect on the other particle – which by
now could be on the other side of the universe. In short, Einstein
and his colleagues had ingeniously discovered a case where
observation could not, apparently, be involved in determining
some fact about a subatomic particle.

Here Einstein rested his case, assuming that he had demonstrated
that there must be some hidden variables, some intrinsic feature
of subatomic particles, to explain the way they behaved – for if
there was not, it would mean that particles that were completely
unconnected could magically influence one another.

In 1965, however, the Belfast-born physicist John Bell at
Europe's CERN laboratory laid the theoretical foundations for just
such a test, one that would prove once and for all whether the
Copenhagen interpretation was right. The EPR paradox assumes
'locality', that things that are remote from one another, that are
not directly linked, cannot effect each other. This was a perfectly
reasonable scientific assumption to make. To believe otherwise is
to believe in the possibility of telekinesis, or that the configuration
of stars billions of miles away can influence the character of a
newborn baby. Bell's theorem showed a way of experimentally
determining whether or not this assumption of locality matched
up to observable results. In 1982 Alain Aspect of the University of
Paris performed the experiment for real and showed that it did

not, that subatomic particles somehow managed to communicate in the way the EPR paradox had assumed to be paradoxical.

It is a mistake to think that Einstein was a 'naive' realist, an innocent believer in the independent existence of a material reality. In a letter to Viscount Samuel, he argued that reality is a feature of the theory used to understand the world, rather than a feature of the world itself. He held this belief for the very good reason (for a scientist) that 'one is in danger of being misled by the illusion that the "real" of our daily experience "exists really", and that certain concepts of physics are "mere ideas" separated from the "real" by an unbridgable gulf'.[18] His objection to quantum mechanics was not its implications for materiality, but the idea that the physical world should be regarded as a wave function, a set of possibilities, that are collapsed into an actuality by observation. It implied that the world was created simply by our perception of it – indeed, it threatened to force science back into the speculative realm of philosophy, specifically of the sort of philosophical idealism pithily expressed in a famous limerick by the theologian Ronald Knox:

> There once was a man who said 'God
> Must think it exceedingly odd
>   If he finds that this tree
>   Continues to be
> When there's no one about in the Quad.'[19]

. . . to which came the anonymously penned reply:

> Dear Sir, Your astonishment's odd:
> *I* am always about in the Quad.
>   And that's why the tree
>   Will continue to be,
> Since observed by Yours faithfully, God.'[20]

Whatever Einstein's misgivings, the reality created by the classical science of the Newtonian era, which Einstein had himself sustained as well as supplanted with his theory of relativity, had gone for ever, at least when it came to explaining the subatomic phenomena that must underlie all events at the scale of the molecule and above. Quantum mechanics, its confirmation in Bell's theorem and in the experiments of Alain Aspect and others, its practical application to the 'real' world, its success in creating a theoretical foundation for the entire edifice of information technology – all of these things swept the classical era and all it represented aside. It threatened, in other words, a 'paradigm shift', the moment described by Thomas Kuhn when the old ideas and practices, enfeebled by anomalies, are forced to make way for new ones.

For the time being, large-scale mechanics and quantum mechanics have been forced to co-exist, because neither is any good at explaining the other. Nevertheless, philosophers and theoretical physicists have struggled mightily to come up with a new paradigm, a more acceptable interpretation of the 'quantum reality' of the subatomic world. It has been this struggle that has alerted the shamans and soothsayers to the dawning of new eras and the transitions of planetary dimensions. If even science is starting to question the nature of reality and our knowledge of it, then something must be up.

* * *

One outcome of quantum mechanics has been a growth in enthusiasm among some scientists to ask the sorts of speculative, not to say imponderable, questions about the nature of reality that had previously been pondered by philosophers. The Copenhagen interpretation is regarded by many of these scientists as a stopgap measure, an unsatisfactory, incomplete, fundamentally agnostic view. There are, however, those who seek to retain it on the grounds that, crazy though it may be, it works.

John Wheeler, director of the Center for Theoretical Physics at Texas University, believes the Copenhagen interpretation to be 'unshakable, unchallengable, undefeatable – it's battle tested'[21], and thinks it is our assumptions about reality, rather than the theory, that must change. At a conference convened to celebrate the centenary of Einstein's birth, he tried to show how we could be expected (literally) to entertain such an idea by recalling a game of twenty questions. He was asked to leave the room while his friends thought up the word he was to identify asking his 20 yes/no questions. His suspicion grew that something was up when he was kept waiting outside for longer than usual.

Nevertheless, on being readmitted, he set about trying to discover the word by the usual means: 'Is it animal?' 'No.' 'Is it mineral?' 'Yes.' And so on. As the game progressed, the inquisitors found it harder and harder to answer his questions. Finally, he made a guess: 'Is it "cloud"?' 'Yes' came the reply and everyone burst out laughing.

His friends' little conceit was that they did not select a word at all, but simply gave whatever answer to the questions they liked. The only condition was that every answer had to be consistent with the previous answers. The moral of Wheeler's story was that, like the naive classical physicists, he had imagined there to be a 'real' answer when in fact only the answers were 'real'. Reality had emerged through investigation.

The sort of interpretation promoted by Wheeler has often been cited by non-scientists in support of the view that reality is no longer the fixed, independent realm the modern era assumed it to be. It seems to confirm Timothy Leary's view that 'realities are determined by whoever determines them',[22] and Jaron Lanier's oft-quoted assertion that it is not the idea of artificial reality that is strange, but reality itself. It also adds weight to the postmodern belief in pluralism, its distaste for the assumption that there can be one way of looking things, one answer to any problem.

For dogged realists, however, Wheeler's game was hardly sufficient. A much more acceptable, though perhaps even more bizarre, theory for them is the 'many worlds interpretation' formulated by Hugh Everett at Princeton in the 1950s. Everett addressed the assumption of Schrödinger's Cat experiment, that the act of observation settled the cat's fate. Would a passing dog that managed to open the chamber's door collapse the wave function, or would it inadvertently become caught up in it until its owner came along? Why assume that a human collapses the wave function – could not the experimenter become trapped in the cat's state of suspended actuality, absorbed into the system until another observer (perhaps an animal rights activist concerned at the cat's fate) happens by, who is also trapped, and so on until God or some other universal observer breaks the perpetual chain of observation? Is our role that of metaphysical voyeur, condensing the world into actuality through the mere act of perceiving it?

The physicist Eugene Wigner argued that it must be human consciousness that ends the chain of observation. Everett could not accept this – why privilege humans? Instead, he proposed that the cat must occupy two almost identical universes, differing only in that in one it lives while in the other it dies. If ever there was a call for Occam's razor to be wielded by Sweeney Todd, this would seem to be the occasion. The many worlds interpretation is the most extravagant of any theory known in its multiplication of entities – every subatomic event implies at least one new universe (or a change in one or some of an infinite number of parallel universes). Nevertheless, Everett argued that it made much more sense of the subatomic realm than the Copenhagen interpretation. If you can swallow the premise, then everything else seems to fall into place. With the EPR 'paradox', for example, the measurement of a particular attribute of one particle does not magically disturb its distant pair. Instead, the decision to measure one attribute is a decision to live in one particular universe rather than its complementary one. All possible states of a quantum system really exist, only they occupy different, parallel universes.

Unfortunately, living as we do at a level where quantum mechanical effects are not experienced, we cannot see into these other universes. For critics of the many worlds interpretation, this demonstrates that the theory is no better at predicting the outcome of subatomic experiments than the Copenhagen interpretation. Indeed, some physicists even object to the idea that there are different 'interpretations': 'There is only one way in which you can understand quantum mechanics,' argued the physicist Sir Rudolf Peierls. 'There are a number of people who are unhappy about this, and are trying to find something else. But nobody has found anything else which is consistent yet, so when [people] refer to the Copenhagen interpretation of the mechanics what [they] really mean is quantum mechanics.'[23] Why bother with all these other universes when there is a perfectly satisfactory, fully-operational framework in place?

For David Deutsch, a theoretical physicist and champion of Everett's theory, the answer is that the many worlds thesis is more than an interpretation, a way of looking at the same facts which is formally identical to the Copenhagen interpretation. It implies that the universe is fundamentally different. In the 'conventional', Copenhagen view, there is no actual universe beyond its observation; in Deutsch's view, there is an objective reality which is divided up between an infinite number of parallel universes. Furthermore, he thinks there is a way of discovering these other universes.

According to the Copenhagen interpretation, the point at which the wave function collapses happens at no precise moment, but has something to do with the time the system in consideration (say, the cat in the chamber) is consciously perceived by the observer. For Deutsch, this transition moment is significant, because that is where the interpretations divide. According to the conventional interpretation, something happens at that moment – the observer collapses the wave function, the cat dies/survives, and the world is changed forever. In Deutsch's view, the outcome of the experiment occurs at the point when the radioactive sample (to continue with the cat example) exercises its 50 per cent chance of emitting a particle – this being an event that happens quite independently of anyone observing it. At that moment, the two possible versions of the experiment exist in two parallel sets of universes. The outcome observed by the experimenter when the chamber is unsealed depends on which of these parallel universes he or she happens to be in.

To test this out, Deutsch imagined building a 'quantum memory', a device that would be capable of observing and recording the outcome of a quantum experiment as a set of quantum states. Since humans do not (or are not known) to have such a memory, Deutsch imagined building an artificial one – a computer, in other words. This computer could then 'observe' the outcome of an experiment. According to Deutsch, what should then happen is that the memory effectively splits into two copies of itself – because, according to the Everett interpretation, the act of observation should have no special influence over the experiment's outcome. These two near identical copies should then be detectable because, like the two copies of the experiment observed, they will display the interference patterns arising from the uncollapsed wave functions. If Everett's interpretation is wrong, there will be no such interference, because the memory will not have split into copies but, quite the opposite, resolved all the possible outcomes into one actual result.

Deutsch's experiment depends, of course, on the development of a quantum memory, a quantum computer. Is it, then possible to build such a machine? There is certainly growing interest in the relevant technology. In 1990, at the University of Sussex, Terry

Clark announced that his research team had succeeded in using a superconducting circuit about one centimetre across called a SQUID to observe an uncollapsed wave function.[24] Clark compared this feat with Perseus stealing a glance at the gorgon Medusa without being turned to stone. In this case, he saw electrons in a SQUID flow in both directions at once – in a superposition of two states. Clark claimed that such devices could form the basis of a new sort of technology, a technology that could exploit the subatomic phenomena of quantum reality on a superatomic scale. Perhaps such technology could form the basis of a quantum computer in the twenty-first century, a machine capable not only of offering an enticing peek into parallel universes, but of navigating such universes with the aim of discovering which would be the best to live in – a machine with even, perhaps, a switch on its front panel marked 'on/off/both'.

Like the philosophical idealism that seems to be echoed in the Copenhagen interpretation, the many worlds view is itself an echo of what in philosophy is known as 'modal realism'. Like the many worlds view, modal realism is a theory just a little too rich for some philosophical palates: 'when modal realism tells you – as it does – that there are uncountable infinities of donkeys and protons and puddles and stars, and of planets very like Earth, and of cities very like Melbourne, and of people very like yourself,' wrote the philosopher David Lewis, 'small wonder you are reluctant to believe it. And if entry into the philosophers' paradise requires that you believe it, small wonder if you find the price too high.'[25] Nevertheless, mathematicians, in their 'paradise' of numbers, accept the existence of other universes as an abstract corollary of set theory, so why not philosophy, Lewis argued.[26]

Lewis claims that other universes exist in the way someone might argue the existence of 'Loch Ness monsters, or Red moles in the CIA, or counterexamples to Fermat's conjecture, or seraphim'.[27] The reason for making the claim is that it is supposed to overcome the perennial philosophical problem of explaining the status of possible but not actual entities, such as, in Bertrand Russell's famous example, the present King of France. There *is* a King of France alive in the twentieth century, in fact an uncountable number of them (perhaps some of them are you), just none in this universe.

Modal realism also helps provide some sort of account for the status of fiction, the 'truth' ascribed to imaginary worlds. In his book *Fictional Worlds* the literary theorist Thomas Pavel cites a passage from Charles Dickens's *The Pickwick Papers*: 'That punctual servant of all work, the sun, had just risen and begun to strike a light on the morning of the Thirteenth of May, one thousand eight hundred and twenty-seven, when Mr Samuel Pickwick burst like another sun from his slumbers, threw open his chamber window, and looked out upon the world beneath. Goswell Street was at his feet, Goswell Street extended on his left; and the opposite side of Goswell Street was over the way.' 'The reader of this passage', commented Pavel, 'experiences two contradictory intuitions: on the one hand he knows well that unlike the sun, whose actual existence is beyond doubt, Mr Pickwick and most of the human beings and states of affairs described in the novel do not and never did exist outside its pages. On the other hand, once Mr Pickwick's fictionality is acknowledged, happenings inside the novel are vividly felt as possessing some sort of reality of their own'.[28] When the book ends, one cannot believe that the characters, nor the world they occupied, end with it. Fictional realms seem to be greater than the books that create them.

So what sort of reality do they have? Could they exist in some sort of possible world, in the modal realist's paradise? According to Brian McHale, a lecturer in 'poetics' at Tel Aviv University (a very postmodern post), this is precisely the sort of question that postmodernist fiction addresses. Modernist fiction was concerned, he argued, with questions of epistemology – ways of knowing and issues of truth – whereas postmodernist fiction is concerned with ontology – ways of being and issues of reality. This 'ontological' perspective is an interpretation; not just a way of writing fiction, but a way of looking at all fiction. One need only think back to Carlos Fuentes (cited by McHale as an example of a postmodernist author) and the 'reality made out of words and paper' in Cervantes' *Don Quixote*.

Such perspectives fit very neatly with the many worlds interpretation, suggesting the same plurality of worlds, the same startling promotion of possibility to actuality. There are obviously differences in what a theoretical physicist, a philosopher and a literary theorist thinks a possible world may be. Deutsch's quantum worlds are not possible but actual, each one corresponding to each possible resolution of a wave function. Humans could, theoretically, travel between them. Lewis's logical worlds are different in that they each offer a unique and fixed set of logical possibilities, and there can be, he insists, no travel between them: there can be no 'logical spaceship'. Pavel's fictional worlds are 'flexible', imprecisely mapped and, of course, it is very easy to travel between them – one simply has to read a few books. Despite these differences, the way they echo each other seems to be strong evidence of a paradigmatic link, that science is reflecting, or initiating, or even validating, a particular perspective on the nature of reality.

Some physicists still believe that there is a single, external universe, one independent of our perception. Indeed, according to David Bohm, that is what every physicist really believes.[29] Bohm, a victim of McCarthyism who escaped to Britain to become professor of theoretical physics at Birkbeck College, London, argues that there must be 'hidden variables'. A wave function is not a way of working out probabilities but a real field, like a magnetic field, shaped by a 'quantum potential'.

The most distinctive feature of Bohm's theory is his notion of an 'implicate' or 'enfolded' order, which he invokes to explain the way distant particles can magically influence each other as shown in the EPR paradox and its subsequent experimental confirmation. Bohm explains the concept by means of a series of examples. It is, he explains, like the pattern on a piece of paper drawn when the paper is folded, and which disappears to form different patterns when the paper is spread out. It is to be seen in a television broadcast, where the visual image is broken up into lines transmitted one at a time, which results in parts of the image which are near to each other when they are displayed being split apart – unfolded – for transmission and brought together again, 'enfolded', for reception. It is to be seen in a drop of culinary dye put into dough which is subsequently kneaded, (en)folded and drawn out over and over again, pulling the colour into a thin, tangled trail weaving through the marbled bread. It is in the enfolding, unfolding relationship between an animal and its gene. It is in the seed that grows into a plant that creates more seeds – and in the way the plant unfolds not just according to the genetic information in its seed, but according to the environment it is growing in.

Bohm's favourite example of implicate order is the hologram, the photographic process for storing and viewing three-dimensional images on a two-dimensional plate, and which is familiar to most of us in the form of novelty badges showing a winking eye and security marks on credit cards. Holograms have one remarkable feature: the entire picture is contained in its every part. If you shatter a hologram, you see a miniature version of the whole image in each shard. This is how the universe presents itself to the scientist's inspection. An image of a quantum system does not have a point-to-point correspondence to what it depicts; rather, it is an unfolded view of an enfolded order.

It has been argued that the concept of the implicate order is, in fact, irrelevant to Bohm's defence of hidden variables. Nevertheless, it provides Bohm's theory with the means of launching a very paradigmatic assault on the 'mechanistic' view of the universe which arose out of Newtonian or classical science. The mechanistic order assumes that the universe is like a machine, which can be divided into entities that are, as Bohm put it, *'outside of each other*, in the sense that they exist independently in different regions of space (and time) and interact through forces that do not bring about any changes in their essential natures'.[30] The mechanistic view has a fetish for measurement, for tearing the enfolded order apart so that it can be compared with some external, arbitrary standard. Einstein's theory of relativity challenged this notion of order. Quantum physics, argues Bohm, totally destroys it, and the strangeness of the Copenhagen interpretation is a result of science's desperation to retain it. Bohm contrasts the mechanistic view with that of the East. 'In the prevailing philosophy in the Orient, the immeasurable (i.e. that which cannot be named, described, or understood through any form of reason) is regarded as the primary reality.'[31] Reality is lost, not revealed, by measurement.

Bohm's idea of the implicate order seems to express perfectly the paradigm shift that is said to be pushing the mechanistic world view off its pedestal. Like Deutsch, whose theory the idea of an implicate order is actually designed to oppose, he seems to be part of the same continental drift of ideas, his holistic thrust, his interest in hidden order, his sympathy with Oriental philosophy resonating with the preoccupations of postmodernism. The same themes recur throughout the 'new physics' and the intellectual movements that cite it as evidence of a changing world view.

In 1975, the one-time research physicist Fritjof Capra published *The Tao of Physics* in which he denounced mechanistic scientific values, arguing that they had not only become scientifically unsustainable but were threatening global catastrophe. He argued that quantum physics demonstrated the weakness of Western materialism, the danger of iron determinism shackled by the causal chain. Physicists, he felt had confirmed – almost despite themselves – the validity of the 'Tao', the Eastern 'way' that offered a 'new vision of reality . . . based on awareness of the essential interrelatedness and interdependence of all phenomena – physical, biological, psychological, social and cultural'.[32]

In 1979, James Lovelock published his 'Gaia' hypothesis. Unlike Capra's book, it was not meant to be a manifesto, though it has been taken by many as the basis of one. Lovelock argued that it was a fully fledged scientific theory, one that properly fulfilled Popper's falsifiability condition. Invoking through its name the Greek idea of Mother Earth, the hypothesis argued that the earth, its lands and seas as well as its plants and animals, is a living organism. The idea was explained to me by Lovelock with reference to the swimming pool at his country cottage (which he calls an 'experimental station', because suppliers are reluctant to deliver fissile materials to a domestic address), set in a 30-acre plot of Devonian paradise.[33] The pool's water was putrid, not the clean blue hue of chlorinated water but the pea green of stagnant slime. Nevertheless, Lovelock claimed that it was clean enough to drink; he even had a certificate from the water authority to prove it, apparently meaning this to be a comment on the hygiene of the water rather than the water authority. 'Any human parasites that got into there would be gobbled up in seconds,' he pointed out as we watched strange aquatic lifeforms stir the fetid ooze. Few people would have given the human much of a chance either.

Lovelock's claim was based on examining natural systems using what he called the 'top-down' view associated with engineering, as distinct from the 'bottom-up' approach adopted by most physicists. According to the bottom-up approach, things need to be taken apart in order to be understood. The top-down approach, in contrast, considers the system as a whole. In the case of swimming pools, this translates into a strict regime of neglect, enabling a benign state of equilibrium to assert itself naturally. Chlorination and filtration would produce a state of fragile stability rather than natural balance, demanding constant maintenance.

According to the Gaia hypothesis, the world is like Lovelock's swimming pool, governed by self-regulating feedback effects. For example, he observes how the earth's atmospheric temperature has, throughout its history, remained steady despite the amount of energy from the sun increasing by a third. He also thinks that the 'greenhouse effect' produced by industrial gases trapping heat in the atmosphere will probably be trivial in Gaian terms, but that the self-regulating mechanisms that will return the temperature to its historical average could make the environment inhospitable to human life (in other words, humanity will not destroy the world, the world will destroy humanity).

Such ideas of self-regulation and equilibrium are themselves now a central part of the science of non-linear dynamics and chaos – concepts such as emergence, strange attractors and self-organization are applauded by their promoters for having the same 'top-down' trajectory. All are part of what is supposed to be a more ecological political mood, the 'caring', anti-mechanistic 1990s, the same great new paradigm, influenced by the same intellectual climate that produced postmodernism.

No one has yet managed to develop a clear meteorology for this intellectual climate, a way of showing how changes in one field influence changes in another. How is it that 'opinion formers' all seem to be forming the same sorts of opinions about widely differing fields at roughly (very roughly – quantum mechanics is the product of a century's scientific work, postmodernism of just a few decades) the same time? What links quantum reality to postmodernism? In some cases it is obvious, as illustrated by the number of novels published in the late 1980s that explicitly credit the 'new science' as an influence. Ian McEwan's *The Child in Time* had a theoretical physicist as a character who declaimed Bohm's implicate order. In the first chapter of *First Light*, entitled 'The uncertainty principle', Peter Ackroyd wrote: 'everything began moving away. Nothing but waves now, their furrows tracking the path of objects which do not exist.'[34] William Boyd's *Brazzaville Beach* was peppered with illusions to chaos and computation.

Outside literature, the link between science and cultural

phenomena can be less clear. N. Katherine Hayles argued that the science of chaos expresses the same values as 'deconstruction', the concept arising out of literary theory that texts have no one meaning, but many, contradictory ones: 'Deconstruction shares with chaos theory the desire to breach the boundaries of classical systems by opening them to a new kind of analysis in which information is created rather than conserved. Delighting in the increased complexity that results from this "scientific" process, both discourses invert traditional priorities: chaos is deemed to be more fecund than order, uncertainty is privileged above predictability, and fragmentation is seen as the reality that arbitrary definitions of closure would deny.'[35] Hayles's explanation as to why there should be similarities between these two 'discourses' is vague. There is unlikely to be a direct link – scientists tend not to peruse journals of literary theory. Rather, Hayles argues that similarities arise 'because of broadly based movements within the culture, which made the deep assumptions underlying the new paradigms thinkable, perhaps inevitable, thoughts. They illustrate how feedback loops among theory, technology and culture develop and expand into complex connections between literature and science which are mediated through the whole cultural matrix.'[36] Her allusion to 'feedback loops', very much a concept from the paradigm she is describing, is instructive. The difficulty is knowing whether these feedback loops reveal significant differences or merely amplify insignificant ones.

Whatever the mechanism of paradigm shifts may be (perhaps even to look for a mechanism is to show a failure to keep up with the paradigms), the era of 'naive realism' is at an end. Quantum mechanics and the uncertainty principle have not invalidated the principle of objective observation – a wave function collapses in exactly the same way irrespective of who observes it. But they have seriously undermined science's faith in an external, material reality.

'In the overthrow of the old world view – a paradigm shift that is dramatically transforming our understanding of reality – the chief casualty is common sense', wrote Paul Davies and John Gribbin. 'Whereas in the Newtonian picture of reality, human senses and intuition proved a good guide, in the abstract wonderland of the new physics it seems that only advanced mathematics can help us to make sense of nature.'[37] I would question that 'common sense' was ever a part of the Newtonian picture of reality – to the non-scientist, ideas of perfect vacuums through which moving objects never slow down are hardly matters of intuition. Galileo was not demonstrating that the mechanistic world view confirmed common sense when he showed that objects of different weight fall to earth at the same speed – he was impressing his public in the power of science by demonstrating precisely the opposite. Nevertheless, there has been a very real, or at least a 'performatively productive', as Lyotard might put it, shift in the foundations of knowledge – and it is clear in which direction. Reality has left the physical world and moved into the virtual one.

Notes

1   John Naisbitt and Patricia Aburdene, *Megatrends 2000*, London: Sidgwick, 1990, p. 1.
2   Fritjof Capra, *The Turning Point*, London: Fontana, 1983, p. 15.
3   George Gilder, *Microcosm*, New York: Simon & Schuster, 1989, p. 17.
4   Paul Davies and John Gribbin, *The Matter Myth*, London: Viking, 1991, p. 7.
5   Viscount Herbert Samuel, *In Search of Reality*, Oxford: Basil Blackwell, 1958, p. 15.
6   'Cold fusion refuses to die', *Mondo 2000*, 2 Summer 1990, p. 16.
7   Frank Close, 'Cold fusion I: the discovery that never was', *New Scientist*, 1752, 19 January 1991, p. 50.
8   Frank Close, *Too Hot to Handle*, London: W. H. Allen, 1991.
9   'And yet it kills', *New Scientist*, 1608, 14 April 1988, p. 17.
10 Phyllida Brown, 'Monkey tests force rethink on AIDS vaccine', *New Scientist*, 1787, 21 September 1991, p. 14.
11 Jean-François Lyotard, *The Postmodern Condition*, trans. Geoff Bennington and Brian Massumi, Manchester: Manchester University Press, 1986, p. 46.
12 Karl Popper, *Unended Quest*, London: Fontana, 1980, p. 36.
13 Werner Heisenberg, *Physics and Philosophy*, London: Penguin, 1989, p. 19.
14 Heisenberg, 1989, p. 29.
15 Heisenberg, 1989, p. 42.
16 Nigel Calder, *Einstein's Universe*, London: BBC, p. 163.
17 In F. David Peat, *Einstein's Moon*, Chicago: Contemporary Books, 1990, p. 76.

18 Samuel, 1958, p. 169.
19 *The Oxford Dictionary of Quotations*, 1979, p. 305.
20 *The Oxford Dictionary of Quotations*, 1979, p. 4.
21 P. C. W. Davies and J. R. Brown (eds) *The Ghost in the Atom*, Cambridge: Cambridge University Press, 1986, p. 60.
22 Timothy Leary, 'Quark of the decade?', *Mondo 2000*, 7, Autumn 1989, p. 54.
23 Davies and Brown, 1986, p. 85.
24 Philip Ball, 'Schrödinger's cat ensnared', *Nature*, 347, 27 September 1990, p. 330.
25 David Lewis, *On the Plurality of Other Worlds*, Oxford: Basil Blackwell, 1986, p. 133.
26 Lewis, 1986, p. 4.
27 Lewis, 1986, p. viii.
28 Thomas G. Pavel, *Fictional Worlds*, Cambridge, Mass.: Harvard, 1986, p. 11.
29 Davies and Brown, 1986, p. 118.
30 David Bohm, *Wholeness and the Implicate Order*, London: Ark, 1983, p. 173.
31 Bohm, 1983, p. 22.
32 Capra, 1983, p. 285.
33 Benjamin Woolley, 'One world, one life', *The Listener*, 8 June 1989, p. 4.
34 Peter Ackroyd, *First Light*, Tunbridge Wells: Abacus, 1990, p. 3.
35 N. Katherine Hayles, *Chaos Bound*, New York: Cornell University Press, 1990, p. 176.
36 Hayles, 1990, p. 4.
37 Davies and Gribbin, 1991, p. 11.

12
DISCOVERY

'We are on the brink', wrote Harvey Rheingold in his book on virtual reality, 'of having the power of creating any experience we desire.'[1] This expresses the position of what one might call the naive virtual realist. For people like Rheingold, the computer is a reality machine without limit, capable of creating any world we desire to live in. This brink: how close is it? When will we have this power, and how will we acquire it?

Early models of the head-mounted display were awkward and heavy, uncomfortable to wear for more than a few minutes. The tracking of the wearer's movements was crude, the resolution and colour of the screens poor, and the rate at which pictures were displayed slow and out of step with the wearer's movements. However, one can safely predict that many of these limitations will be overcome as the technology develops. All the major hardware components have the potential for improvement. One can imagine in, perhaps, a decade or so a lightweight headset no more cumbersome than a pair of Walkman headphones and spectacles, perhaps connected to a Walkman-sized portable computer.

One can even imagine, as Howard Rheingold did so enthusiastically in an article in *Mondo 2000* and subsequently in his book, the development of 'teledildonics'. The term 'dildonics' was coined by Ted Nelson to describe a machine that converted sound into tactile sensations. The 'tele' was added later to denote the communication of such sensations over a distance. 'Picture yourself in a couple [of] decades hence, getting dressed for a hot night in the virtual village', speculated Rheingold. 'Before you climb into a suitably padded chamber and put on your head-mounted display, you slip into a lightweight – eventually, one would hope diaphanous – bodysuit. It would be something like a body stocking, but with all the intimate snugness of a condom.'[2] This wraparound prophylactic would be embedded with thousands of tiny effectors which, under the computer's control, would simulate the feel of any object and material from satin to skin. You then plug it into the phone network, connecting to someone else wearing a similar stocking, and then get on with the business of erotic interfacing.

The sleazy mix of telephone sex and an updated version of Woody Allen's 'Orgasmatron' offered by the teledildonic scenario was obviously an AIDS-inspired idea, but, at the time of writing this book, the technology of the sensory body stocking was not even a distant possibility. Nor was the even more ambitious idea discussed by virtual reality developers of dispensing with prosthetic interfaces altogether and wiring the computer straight into the brain.

Nevertheless, all of these things are in principle possible. But what of the information they will deliver? What of the virtual reality generated by the computer that is experienced through these new, intimate interfaces? How realistic will it be?

In the early years of virtual reality, the best simulations were produced using 'graphics workstations', desktop systems specifically designed to generate graphics. These were among the most advanced computers around, but nevertheless comparatively cheap, well within the resources of small development companies and university departments. Such computers could only manage a primary-coloured virtual landscape made up of simple geometric shapes, and could only update the picture seen through the head-mounted display a few times per second, producing a lag between the wearer's movement's and the corresponding changes in the image. 'Photorealistic' images – the goal of most computer graphics developers – can be produced, even using personal computers, but the generation of a single image can take hours, even days.

The problem is related to the amount of processing that is required to simulate a three-dimensional space and the objects it contains. The space itself, and the objects within it, are created only once, as a series of mathematical descriptions or 'models'. These objects are then 'rendered' into a visible image by calculating how they will appear to the user from the current point of view (taking into account light sources, the position of objects in front of and behind the object concerned, shadows and so on). It is this rendering process of turning the mathematical description into a viewable (and, as the technology develops towards Sutherland's ultimate display, a tactile, audible, even smellable) scene, that is usually done in 'real time', as each image is required. Suppose, for example, that you wanted to experience standing at the foot of Mount Everest. The computer would hold a pre-defined model of the mountain describing its geometry. As you looked up, the computer would use the information from the model and from the tracking device attached to your head-mounted display to generate the picture of Everest as it would be seen from your position, the size and perspective of the mountain side being determined by your distance from it.

Such a task requires an enormous amount of computing power. A high-resolution computer-generated landscape scene is made up of millions of basic geometrical shapes ("polygons"), the exact shape of each having to be adjusted as the point of view changes. Not even the most powerful supercomputers are capable of performing the trick photorealistically in real time. The question is: will they be able to do so in the future? In a project entitled 'Micro 2000', engineers at Intel, the company that developed the first microprocessor, predicted that by the turn of the century, microprocessors would be capable of two billion instructions per second, making them over a thousand times faster than current desktop computer microprocessors.[3] With such formidable power available, computers would be capable of creating relatively rich imagery in real time. But would they be photorealistic?

The computing industry's fetish for power and speed has distracted it from examining what it means by 'photorealism'. Most regard it as a product of higher resolution and more colours. However, a grainy, black and white picture can look more 'photorealistic' than a TV-quality full-colour computer image. The reason seems to be that the realism lies in the image's content, not the quality of its reproduction. A photorealistic image is one that looks as though whatever it depicts is in some sense real – it concerns, in other words, the sophistication of the computer models, of the descriptions of the virtual objects and landscapes, as much as their rendering.

Most of the effort to develop modelling techniques has concentrated on creating 'computer aided design' (CAD) tools, which are used by designers to create two- and three-dimensional engineering models of products. These are fine for prototyping purposes, but not much good at 'creating any experience we desire'. Techniques have been developed which help produce more 'natural' forms: fractal geometry, for example, is effective at simulating the shape of mountains, vegetation and clouds; simple objects can be given a complex shape through 'texture mapping', where a three-dimensional surface description can be 'mapped' on to a basic object (such as a tread texture onto a doughnut shape to create a tyre); 'genetic' procedures can be used to nurture new forms, with shapes being 'grown' from a 'gene' in a computer-generated environment, where artificial selection picks out the more attractive or suitable mutations. But using even these techniques, creating models of complex virtual environments remains a formidable task, one that is not helped by the current practice of creating models on a one-off basis.

In the late 1980s, companies started selling libraries of simple shapes to users of AutoCAD, the industry-standard personal computer CAD package developed by one of the promoters of virtual reality technology, Autodesk. Designers would buy these 'clip objects', as they came to be known, to adapt and integrate into their own designs. One company even produced a CAD mannequin, a simple model of a human form, which could be used to 'test' ergonomic factors in the design of furniture and workspaces. If this market were to develop further, investment funds may become available for developing existing models – perhaps even whole landscapes. The American artist Matt Mullican has already built a 'metaphysical city' on a powerful Connection Machine supercomputer. Perhaps film producers will one day build such a city to create a convenient, controllable setting for urban movies. Perhaps a new profession will emerge, that of hyperreal estate agency, engaged in the trading of models of simulated environments for virtual reality users to explore.

Creating landscapes and inanimate objects, which are fairly static, is one thing. Creating creatures is quite another. In 1988, the MIT Media Lab started showing its animation short 'Cootie gets scared' at computer graphics festivals to demonstrate its development of 'smart' animated characters that knew how to walk and respond to simple 'stimuli' encountered in the virtual world of the computer animation. In the case of 'Cootie', the character was a cockroach scared by a spider – it was no Mickey Mouse, but it demonstrated the idea impressively enough. The previous year, the American computer graphics company Symbolics had shown an ambitious animation entitled 'Breaking the ice' featuring birds and fish which were claimed to be individually programmed with the same flocking instincts as sperm. Such experiments have shown that very primitive behaviours can be simulated, laying the foundation, developers hope, for higher forms of life to be reproduced.

Rheingold is, I feel, optimistic in thinking that these ideas and developments will come together to deliver us to the brink of having the power of creating any experience we desire. Computer-generated pictures will certainly become more complex and colourful. They will increasingly dominate our visual culture as they become cheaper to produce than artificial images created using conventional means. But the idea that, for example, virtual reality systems will imminently be capable of realizing a fictional world indistinguishable from the real one seems to skirt round some unanswered questions about the nature of fiction and the nature of simulation. Is the computer just a new creative medium? Are fictional worlds the same as virtual worlds? Do computers create reality, or discover it?

* * *

According to the BBC's authoritative history of computing, *The Dream Machine*, the computer is fundamentally different to other sorts of machines because it is so adaptable. 'The computer is not so much a machine as a powerful personal medium of expression and communication,' wrote the series' authors in the accompanying book.[4] All sorts of implications follow from this view of the computer. For example, it encourages the view that psychology will reveal the computer's capabilities. A lot of the early research into interfaces was based on studies of child behaviour. Children, argued the Media Lab's Seymour Papert, 'give us a window into the way the mind really works, because they're open . . . I think we understand ourselves best by looking at children'.[5] As the prayer-book says, out of the mouths of babes and sucklings . . .

The experiment most often cited as an example of how children could help untap creativity was performed by the Swiss psychologist Jean Piaget. He asked children of about five to pour the same amount of liquid into two glasses, one tall and the other short, and then say if the glasses contained the same amount of liquid. Piaget observed that the children would assume that the tall glass contained more, and this, he argued, demonstrated how the visual mentality dominates the intellectual one. For this reason, graphical interfaces are regarded as more 'intuitive' than text-based ones, because the child-like responses that we first apply to solving new problems rely on visual stimuli rather than symbolic reasoning.

A post-infantile part of me is suspicious of this emphasis on child psychology – the child, after all, was wrong about the amount of milk in the tall glass. There is, perhaps, a hint of PR in the enthusiasm among computer companies to have their products used as the basis of educational research projects. The industry's greatest fear is that sceptical consumers will find their products too complex and lacking in any obvious productive purpose. What better way, then, of making them 'user friendly' than by associating them with sweet, innocent children, and what better purpose to give them than some indistinct power to educate? Papert told Stewart Brand that kids like computers because 'they can feel the flexibility of the computer and its power. They can find a rich intellectual activity with which to fall in love.'[6] Perhaps, but kids can find a pretty rich intellectual activity with which to fall in love using wood blocks, too. It does not seem to me obvious that children have a some sort of mystical affinity with computers simply because a lot of adults find them intimidating. It is possible the adults are expressing a legitimate fear of technological domination that children have not got the conceptual equipment to identify. Masters, as we know, very often disguise themselves as servants.

Howard Rheingold claims that the cave paintings at Lascaux in France were 'primitive but effective cyberspaces', because they were used to create a 'three-dimensional sound and light show' that was possibly used to 'imprint information on the minds of the first technologists'.[7] 'Computers are theater', wrote Brenda Laurel. 'Interactive technology, like drama, provides a platform for representing coherent realities. . . . Two thousand years of dramatic theory and practice have been devoted to an end which is remarkably similar to that of the fledgling discipline of human-computer interaction design; namely, creating artificial realities in which the potential for action is cognitively, emotionally, and aesthetically enhanced.'[8]

Laurel and Rheingold are expressing the very paradigmatic view that reality is a cultural artefact, and that the computer is a way of enriching it, giving more people more control over its construction. Virtual realities are real by virtue of our interaction with them, rather than by virtue of what they are. Therefore, it is psychology, the 'physics of virtual reality', as William Bricken called it, that determines what is and what is not virtually 'real'. Most virtual realists are quite content with this, since it provides them with a scientific strategy for researching VR. They seem untroubled by the fact that, as Gregory MacNicol wrote in a sceptical article about virtual reality, 'at present, we simply don't know enough about how people think and about how the sensory processes interact with the emotions'.[9] Indeed, it seems foolhardy on available evidence to thing that 'we' are *ever* likely to know 'how people think and about how the sensory processes interact with the emotions'.

James Boswell, in his famous biography of Samuel Johnson, relates a discussion with Johnson about the idealism of Bishop Berkeley, which proposed the non-existence of the material world. Boswell pointed out that though it may not be true, it was impossible to refute. 'I refute it *thus*', retorted Johnson, kicking a large stone.[10] Outside the field of paradigms and postmodernism, I would imagine that most people would think of themselves having Dr Johnson's robustly uncomplicated attitude to reality. Few think it the slightest bit odd to think that the tree continues to be when there's no one about in the quad. Baudrillard may assert that the world is just a simulation and that the Gulf War Did Not Take Place, but no one seriously believes him. Some have argued that he is using sophisticated arguments to avoid confronting very real moral and political issues.[11] Quantum reality may be strange, but it simply has no relevance to the world most of us experience. I explained to an acquaintance that, unobserved, the sofa he was sitting on was smeared across the universe in a field of unactualized possibilities. 'Really?' he said, looking at it admiringly. 'Where did you get it from?'

When I first donned a virtual reality rig, the last thing I thought of was reality. It was uncomfortable and the images I saw before my eyes made little sense. Nevertheless, I found it exciting – not intuitively, but intellectually. The objects I was seeing seemed to have some sort of independent existence that the shortcomings of the technology were only allowing me dimly to experience. The computer generated images depicted some sort of 'reality' because they were the product of simulation, of a mechanical, algorithmic procedure. They were generated by the application of laws based on mathematical principles. The crude, primary-coloured wendy houses I explored in virtual reality demonstrations were three-dimensional geometric models that were formally identical to physical buildings. What they looked like from a particular point of view was a matter of mechanical computation, not human imagination, either on my part or that of the system's designer.

At the computer graphics conference described by Timothy Leary as the most important ever held by human beings, the respected computer industry analyst Esther Dyson counselled the audience not to forget symbols. It was interesting advice. In semiotics, a distinction is drawn between three forms of repre-sentation: iconic, indexical, symbolic. An icon (a word that has a different, semi-technical meaning in computing) has some neces-sary, formal relationship to what it represents. So, a photograph of a rose is iconic because the shape of the image subject bears some necessary relation to the rose it depicts. An index is a sign that bears a causal or sequential relationship to what it represents. So, smoke is the index of fire. A symbol has only an arbitrary relationship with what it represents. The word 'rose' bears no necessary or causal relationship with a rose. Its meaning, according to the structuralists, comes from its 'structural' relation-ship with other words.

Dyson was defending the unique power of symbols just as, implicity, Baudelaire did with his attack on the public's obsession with the 'truth' of photography, because symbols are necessary to knowledge. Being arbitrary, they can be used to make abstract generalizations and observations. They are the tools of human imagination and creativity. David Hockney welcomed the intro-duction of computer manipulation techniques to photography because, as he put it in a television interview, 'we've tended to think of the photograph as the ultimate representation of reality on a flat surface, and I think that's naive, frankly. [Photography] is losing its veracity. After all, people have assumed drawing and painting never had this anyway. But they're going to realize photography doesn't either.'[12]

In their analysis of the postmodern condition, cultural theorists and structuralists have tended to regard the symbol as the dominant form of representation. Everything, even things that we might not have thought of even as representations, is a 'text' – a film, a building, a city and, ultimately, reality.

It is true that we are awash with symbolic representations. Even quite conventional photographic images are, by virtue of the way they are composed and what they really 'mean', only incidentally iconic. To return to an example used earlier, a film of a group of people on a desert island consuming a chocolate bar is first and foremost a symbolic representation of whatever qualities the chocolate bar manufacturer wants to ascribe to its product – the iconic qualities that come from it being a film showing 'real' people on a 'real' island eating a 'real' bar serve simply to persuade us that there is something real about what the advertising is saying about the bar.

The computer's apparent universality has distracted us from the important fact that it performs just one function; it computes. All it can do is perform a series of algorithmic, mechanical operations. The dynamics of a simulation, and the interaction that arises out of them, are the product of mechanical operations, not human imagination. It is, to this extent, iconic. This is what makes it realistic, just as the mechanical photographic process made photographs realistic. The problem is working out to what degree it is also symbolic. Whoever described television as the 'window on the world' failed to point out that the window is small and the world big; there is only so much that can be shown at any one time, which is why there are television producers and directors who select what is included and excluded. A flight simulation is a symbolic artefact, too, to the extent that the landscape it simulates, even if it is based on a real landscape, is the product of human design.

The postmodernist paradigm, with its preoccupation with narratives and texts, has tended to treat mathematics as just another symbolic system of representation. All reality, it implies, is a construction of language – including the reality revealed by mathematics. From a scientific perspective, the two are fundamentally different: one is the language of humans, the other, to paraphrase Galileo, the language of nature. Maths is not simply a form of representation, at least not of the physical world, as the work of Hilbert, Turing and Gödel sought to show. As observed in an earlier chapter, the mathematician and writer Andrew Hodges emphasized the difference in his dismissal of an argument as being 'only words' – unreal and therefore ultimately trivial because language can be used to say anything. The paradigm shift in science has not in any way undermined this faith in mathematics – if anything, it has reinforced it. Quantum mechanics relies on a mathematical expression, the wave function, as the ultimate description of the universe.

Perhaps we should not be surprised that postmodernists and critical theorists see reality as residing in language, and scientists see it as revealed by mathematics, because those are their chosen areas of study. What is surprising is that the computer industry should have tried to straddle the two, and that it should expect to bring about a happy union of the two in virtual reality. There is a hope underlying all the speculations about the future of virtual reality that it will use mathematics to explore the realm of the human imagination. The idea of the interactive movie, to take just one example, is based on the assumption that whatever is plotted in a script is in some important way computable; that there is a phase space, to use mathematical terminology, of possible plots that can be modelled using a computer and explored using a virtual reality rig. The question this begs, namely whether the realm of the human imagination is computable, the virtual realists assume to have been answered. Indeed, the matter seemed to be considered closed in the computing industry when Marvin Minsky came up with the term 'artificial intelligence'. Having accepted the hard computationalist position that all of reality is computable, capable of simulation, the virtual realists have then gone on to promote simulation as a way of enhancing human creativity.

In order to make sense of this position, one has to look at its origins within the computing industry itself. Computing is both a technology and a science. The technologists are concerned with developing the computer as a practical tool, the scientists with developing it as a tool of discovery. In his landmark paper on the 'Ultimate display', Ivan Sutherland saw his hypothetical device as a 'looking glass into a mathematical wonderland'. Computational science is concerned with discovering the limits of that wonderland, in answering the question: 'What can be mathematically modelled?' There is no reason why non-scientists should want to explore this mathematical wonderland, so the technologists changed the agenda by persuasively arguing that they could use the 'ultimate display' as a way of exploring the wonderland of the human imagination, as a way of making such a wonderland more real.

Although I have expressed some scepticism at the idea that virtual reality can reveal this wonderland, the technologists have opened up what may prove to be a significant beach-head on humanist territory. The 'logic' of narrative remains a perplexing mystery – why is it, for example, that a naturalistic film about domestic life can seem less convincing, less 'real', than a fantasy about mutating robots? It is possible to imagine that a good story works because it is closer to some ideal, some truth, that we dimly perceive through our imagination. However, this intuition is not sufficient to prove that even if this ideal does exist that it is computable. This is not to say the computer's use as an expressive medium cannot ultimately help settle the issue. Maybe Ted Nelson's Xanadu will reveal important features about the relationship of different literary artefacts. Maybe a method will one day be developed that encodes fictional worlds as databases that can then be explored using the computer – perhaps via a language-based or symbolic interface (yielding interactive novels) or via a geometry-based or iconic interface (virtual reality). But no one has yet produced any evidence that this will be the case. Building more intimate interfaces, generating more photorealistic graphics, examining more psychological experiments is not enough to settle the matter one way or the other. They do not, in other words, show that fiction is a form of virtual reality. They do show that the computer does indeed have interesting and possibly productive uses as a 'creative tool' – it is very useful for manipulating text, as any word processing enthusiast will attest; it can mix media in interesting new ways; it can speed up the process of creating and manipulating images; it can produce convincingly professional documents; it can generate persuasive animations – they may be more compelling, more informative, more useful, but none of these things are any the more 'real' for being processed by computer, and the rhetoric of reality should not be used to lead people into thinking otherwise. To ascribe the computer as a medium with powers to 'discover' the imaginary realm would be as absurd as ascribing the typewriter with the power to discover the world of literature.

The scientific perspective of the computer as a 'looking glass into a mathematical universe' is, however, different. Here, it does have some claim to revealing a reality that was previously hidden. In 1990, NASA's two exploration probes, Voyager 1 and Voyager 2, floated off into deep space having completed their decade-long mission to scan the surface of the outer planets, Jupiter, Saturn, Uranus and Neptune. It was the most successful project undertaken by the agency since the Apollo 11 landed on the moon on 21 July 1969, a happy distraction from the Shuttle's ill-starred deployment and growing public indifference to the idea of deep space travel. In 1986, NASA reached its nadir with the Challenger disaster. Richard Feynmann, member of the enquiry team, sternly reminded NASA's managment that 'reality must take precedence over public relations, for Nature cannot be fooled.'[13] It was a telling choice of words. NASA had been engaged in one of the great scientific adventures. It had helped prove that space, the celestial realm, was part of the same universe, part of the same reality, as the earth. It had helped confirm science's extraordinary predictive powers – for prior to the Soviet Union successfully launched Sputnik 1, there was no reason other than scientific theory to assume that space did not collapse beyond the clouds.

The Voyager probes sent back so much data that NASA's scientists were beginning to be overwhelmed by it. They had received glimpses of the outer planets, but little more understanding. At the Jet Propulsion Laboratory (JPL) in Pasadena, California, however, they started to look at ways of using the computer to 'visualize' all this undigested data. The result was a series of detailed, realistic and beautiful images of the planets. What was interesting and important about these images was that they showed what could not have otherwise be seen. For example, even though the probes scanned the planets from a distance no closer than 26,000 miles, by consolidating and processing all the information gathered, combining views from different angles and scales, the JPL scientists were able to produce animations that simulated what a spaceship-mounted camera would show flying just a mile or so above the planets' surface. They even claimed that the computer-generated pictures of Jupiter's Great Red Spot were sufficiently detailed to validate a mathematical model that explained its formation.[14]

NASA was, of course, one of the most important sponsors of virtual reality technology, through the research work undertaken by Michael McGreevy, Scott Fisher and others at its Human Factors Research Division of the massive Ames laboratory in California. The Ames research clearly has relevance to the visualizing work at JPL. It is quite possible to imagine that a virtual reality rig could provide a new breed of virtual space explorer with the interface they need to inspect such visualizations – perhaps even to walk the surface of a planet simulated by reconstructing detailed data sent back from deep space by probes like the Voyagers. Though such visualization does entail some exercise of the scientists' imagination, mainly to fill in gaps in the data and in the use of 'false colour' to highlight significant features, and though there may be errors in the data or the way it is processed, such visualizations have a strong claim to representing some sort of reality, because they are the algorithmic product of a mathematical model; they are generated by processes that are ultimately independent of human intervention.

One of the great technological hopes among experimental scientists of the late 1980s was the creation of a 'teraflop' computer, a computer capable of calculating a million million sums per second. At the time of writing this book, Thinking Machines was already claiming that it had designed, though not actually built, such a machine,[15] and most other so-called supercomputer manufacturers anticipated achieving equivalent speeds by the end of the decade. Such monstrous mathematical engines would, as *Scientific American* put it, 'become the cross-disciplinary equals of the superconducting supercollider or the scanning tunnelling microscope – portals to new insights and questions in virtually every scientific field'.[16] One could argue that this is simply another excursion in science's mission to discover ever more expensive instrumentation to make ever less relevant observations. But such a machine would be concerned with a much wider range of phenomena than the current technology can cope with – from world climate patterns to the complex chemistry of genetic expression.

One of the most interesting animations to reach the computer graphics film festival circuit in 1990 was a study of the evolution of a storm, produced by the atmospheric scientist Robert B. Wilhelmson at the University of Illinois's National Center for Supercomputing Applications. The storm was depicted as a swelling translucent cloud travelling over a flat surface covered with tracers – arrows and little balls – which revealed the swooping, spiralling airflows within the storm. It took Wilhelmson a year to compute the model using one of the world's most powerful supercomputers, a Cray 2. The usefulness of the result may not of itself have justified the effort, but it very clearly demonstrated how visualization techniques, because they are synthetic rather than analytic, can reveal patterns and structures within a moving system that could never be discovered by examining the static data that produced it.

The effectiveness of a synthetic, top-down approach to experimentation was demonstrated in the nineteenth century when astronomers tried to make sense of the erratic orbit of Uranus. Rather than just ponder Uranus's path in isolation, researchers calculated what the path of such a planet at such a distance from the sun would be if another, hypothetical planet was placed at an orbit beyond Uranus. They then adjusted the size and distance of this hypothetical planet to see if the orbit of the planet representing Uranus could be made to conform to that of the real Uranus. Having discovered that it could, they returned to their telescopes and discovered Neptune. The computer is perfectly suited to performing such experiments, because it can simulate systems such as the solar system, and show what would happen when various parameters are changed.

The science of chaos and other aspects of what are known as non-linear dynamics have shown that beneath the sometimes inscrutable noise of everyday experience there are hidden patterns, a deep structure. In its effort to keep its findings uncontaminated by culturally and socially determined factors – in other words, phenomena that are the result of human free will rather than causation – science has in many respects been driven to the margins of reality, into the centre of the molecule and out to the edges of the universe. A more synthetic approach can afford to be less fussy, because it does not rely on unpicking the chain of cause and effect so much as discovering significant patterns. Though the more traditional scientists remain sceptical that such patterns are to be found in social or cultural phenomena, the computer explorers are more optimistic.

We tend to think of human artefacts as existing independently of any reality. The value of money, for example, is the result of our choice to treat it as having value, it is not intrinsic. However, you could argue that it gets its value from the economy, from a system that determines its value accoding to the principles of supply and demand. Printing more money does not automatically enrich the nation, whatever its inhabitants' attitude. The currency's worth is, rather, a product of complex interactions in the global economy, and it is on this economy that its value ultimately rests. The treasuries of most advanced nations now use mathematical models – usually, because of their complexity, implemented on a computer – to work out the effect of the exchequer's decisions. Such models have proved themselves to be pretty inaccurate, but then they are trying to model a system of immense complexity, every bit as complex as the climate. They are also using modelling techniques that were never designed for the era of high powered computing, and which cannot deal with the amounts of data and levels of complexity that a supercomputer can handle. It therefore seems quite conceivable that there may be a 'reality' to economic affairs that the computer can discover, a pattern that will make predictions about the future of a country's economic welfare far easier to make, and therefore make actions to shape that future in the right direction more effective. More ambitiously, one might hope that the same can be done for other social phenomena, searching for correlations, for example, between the geographic distribution of wealth and crime.

At the level of public relations, the Tokyo metropolitan government commissioned a computer 'simulation' of Tokyo from an advertising agency for display in the foyer of its new city hall. The resulting 'city data map' was a graphic depiction of the city's impressive demographic statistics, showing how the population varied through history (the drop during the war years was a chilling illustration of the ferocity of American bombing raids), land development, property prices, commuter movements. Part of the reason for commissioning the data map was to make up for

Tokyo's lack of identity, for the fact that its fabric is in such a frenzied state of development that, unlike London, Paris or New York, it has failed to establish any sort of stable image of itself. The data map was an image that the government felt appropriate to a city with aspirations to become the capital of the global village. As such, it was no more than expensive PR. But it at least provided a glimpse of the patterns that exist in that great city's perplexing processes, patterns that may ultimately reveal the reasons for its enormous power and success.

In 1989, the American software house Maxis published a best-selling computer game called Sim City. The player's role was that of mayor of a range of possible cities, some based on real cities, some purely fictional. Presented with a map of the city of their choice, and a set of tools for governing and developing it (changing tax rates, constructing roads, demolishing buildings), the mayor's job was to keep his or her popularity rating as high as possible by keeping crime low, growth high and so on. The simulator 'engine' at the heart of the game worked out the consequences of the mayor's choice of actions. If, for example, there was a low density of police stations, crime would rise; if there were too few roads, areas zoned for commercial development would remain undeveloped. Though the engine was relatively crude, being designed to run on a standard personal computer, the game itself was not. It very realistically reproduced some of the patterns of urban development, and demonstrated how difficult it can be to control such development without stifling it on the one hand or letting it run out of control on the other. There were, of course, assumptions built into the model (some of them very political, such as the very low tolerance of the population to even moderate tax rates), but there are assumptions built into any model, scientific or otherwise – it is the business of experimentation to see if such assumptions conform to observation. The result was still an intriguing glimpse of the convincing social phenomena that could be simulated using a relatively simple computer model. Could this be evidence that reality is not confined to the human-free 'nature' studied by conventional science, that there are mathematical laws that determine aspects of social and cultural life as well?

* * *

It is said that Queen Elizabeth of the UK thinks the world smells of wet paint, because wherever she goes, a team of decorators will be round the corner putting a fresh coat on any wall she is likely to pass. I imagined as a child that my world was created much like this, only the decorators was not simply applying new paint, but, under the direction of God, building the world itself, a false reality of sets and props. Around any corner I approached I imagined them to be hurriedly constructing the scenery I would soon behold, while behind me, they were busily dismantling what I had just seen. I imagined that one day I would take an unexpected turn, and catch them out.

But I never have. Instead, I have discovered that it is humans, not God, who are behind this illusion. We are in the midst of artificial reality, surrounded by the constructions of commerce and culture. But is that it? Is there only artificial reality, or is natural reality still there, holding up the flats and drapes of human artifice?

For virtual realists and postmodernists, that, indeed, is it. There is no reality. The gloomy postmodernists think that rationalist values of truth and certainty are being destroyed. The upbeat, bullshitting virtual realists think they can make something better, that dreams can truly be turned into reality. Either way, there is no longer any world independent of our experience of it. We can make it up as we go along.

But is reality really dead? Are we obliged to leave it to the virtual realists to create another for us. Must we sink into postmodern autism? One of the aims of this book has been to show that reality is still there, though not in the material realm of the physical universe where the modern era assumed it to be. In my attempt to distinguish between simulation and imitation, the virtual and the artificial, I have tried to provide a glimpse of where that reality may be, in the formal, abstract domain revealed by mathematics and computation. This is not to say that only mathematics can discover it. Rather, the computer has, through its simulative powers, provided what I regard as reassuring evidence that it is still there.

Notes

1   Howard Rheingold, *Virtual Reality*, London: Secker, 1991, p. 386.
2   Howard Rheingold, 'Teledoldonics: reach out and touch someone' *Mondo 2000*, 2, Summer 1990, p. 52.
3   Kenneth M. Sheldon, 'Micro 2000', *Byte*, 16 (4) April 1991, p. 132.
4   Jon Palfreman and Doron Swade, *The Dream Machine*, London: BBC, 1991, p. 186.
5   In Palfreman and Swade, 1991, p. 103.
6   Stewart Brand, *The Media Lab*, London: Penguin, 1989, p. 123.
7   Rheingold, 1991, p. 379.
8   In Rheingold, 1991, p. 286.
9   Gregory MacNicol, 'What's wrong with reality?', *Computer Graphics World*, 13 (11) November 1990, p. 104.
10 From *The Oxford Dictionary of Quotations*, 1979, p. 275.
11 See Christopher Norris, *Uncritical Theory*, London: Lawrence & Wishart, 1992.
12 *Late Show*, BBC2 11.15 pm, 4 June 1991.
13 In Benjamin Woolley, 'Down to earth', *The Listener*, 120 (3083), 6 October 1988, p. 4.
14 Greg Freiherr, 'Rediscovering the outer planets', *Computer Graphics World*, 14 (7) July 1991, p. 40.
15 Elisabeth Geake, 'Speedy computers hit the price barrier', *New Scientist*, 132 (1797) 30 November 1991, p. 27.
16 Elizabeth Corcoran, 'Calculating reality', *Scientific American*, 264 (1) January 1991, p. 75.

SELECT BIBLIOGRAPHY

Abbott, Edwin A., *Flatland*, London: Penguin, 1986.

Barnaby, Frank, *The Automated Battlefield: new technology in modern warfare*, Oxford: Oxford University Press, 1987.

Barthes, Roland, *Mythologies*, London: Paladin, 1973.

Baudrillard, Jean, *Selected Writings*, Mark Poster (ed.), Cambridge: Polity Press, 1989.

Baudrillard, Jean, *La Guerre du Golfe n'a pas eu Lieu*, Paris: Galilée, 1991.

Berman, Marshall, *All That Is Solid Melts Into Air: the experience of modernity*, London: Verso, 1983.

Bohm, David, *Wholeness and the Implicate Order*, London: Ark, 1983.

Borges, Jorge Luis, *Labyrinths*, London: Penguin, 1970.

Brand, Stewart, *The Media Lab: inventing the future at MIT*, London: Penguin, 1989.

Casti, John L., *Paradigms Lost: images of man in the mirror of science*, London: Scribners, 1990.

Chomsky, Naom, *Syntactic Structures*, The Hague: Mouton, 1957.

Davies, Paul and J. R. Brown (eds) *The Ghost in the Atom*, Cambridge: Cambridge University Press,, 1986.

Davies, Paul and John Gribbin, *The Matter Myth*, London: Viking, 1991.

Dawkins, Richard, *The Extended Phenotype*, Oxford: Oxford University Press, 1982.

Dawkins, Richard, *The Selfish Gene*, 2nd edn, Oxford: Oxford University Press, 1989.

DeLamarter, Richard Thomas, *Big Blue: IBM's use and abuse of power*, London: Macmillan, 1987.

Dewdney, A. K., *The Planiverse*, London: Pan, 1984.

Dreyfus, Hubert, *What Computers Can't Do: the limits of artificial intelligence*, New York: Harper & Row, 1979.

Eames, Charles and Ray, *A Computer Perspective*, Cambridge, Mass.: Harvard University Press, 1990.

Eco, Umberto, *Travels in Hyperreality*, London: Picador, 1987.

Feyerabend, Paul, *Against Method*, London: Verso, 1978.

Foster, Hal (ed.) *Postmodern Culture*, London: Pluto, 1983.

Gibson, William, *Neuromancer*, London: Grafton, 1986.

Gilder, George, *Microcosm: the quantum revolution in economics and technology*, New York: Simon & Schuster, 1989.

Gleick, James, *Chaos*, London: Heinemann, 1988.

Habermas, Jürgen, *Legitimation Crisis*, trans. Thomas McCarthy, London: Heinemann, 1976.

Harvey, David, *The Condition of Postmodernity*, Oxford: Basil Blackwell, 1989.

Hayles, N. Katherine, *Chaos Bound*, New York: Cornell University Press, 1990.

Heisenberg, Werner, *Physics and Philosophy*, London: Penguin, 1989.

Hodges, Andrew, *Alan Turing: the enigma of intelligence*, London: Unwin Paperbacks, 1985.

Hofstadter, Douglas R., *Gödel, Escher, Bach: the eternal golden braid*, London: Penguin, 1980.

Hollingdale, Stuart, *Makers of Mathematics*, London: Penguin, 1989.

Jacob, François, *The Logic of Life*, London: Penguin, 1989.

Jencks, Charles, *The Language of Post-Modern Architecture*, 6th edn, London: Academy Editions, 1991.

Kaplan, E. Ann (ed.) *Postmodernism and its Discontents: theories, practices*, London: Verso, 1988.

Krueger, Myron, *Artificial Reality*, Reading, Mass.: Addison-Wesley, 1983.

Larsen, Judith K. and Everett M. Rogers, *Silicon Valley Fever*, London: Unwin, 1986.

Lévi-Strauss, Claude, *Structural Anthropology*, London: Penguin, 1972.

Levy, Steven, *Hackers: heroes of the computer revolution*, New York: Bantam, 1984.

Lewis, David, *On the Plurality of Other Worlds*, Oxford: Basil Blackwell, 1986.

Lyotard, Jean-François, *The Postmodern Condition: A Report on Knowledge* trans. Geoff Bennington and Brian Massumi, Manchester: Manchester University Press, 1986.

McHale, Brian, *Postmodernist Fiction*, London: Routledge, 1989.

Monk, Ray, *Ludwig Wittgenstein: the duty of genius*, London: Cape, 1990.

Moore, Walter Schrödinger, *Cambridge: Cambridge University Press, 1989*.

Moravec, Hans, *Mind Children: the future of robot and human intelligence*, Cambridge, Mass.: Harvard University Press, 1988.

Morris-Suzuki, Tessa, *Beyond Computopia: information, automation and democracy in Japan*, London: Kegan Paul, 1988.

Nelson, Ted, *Computer Lib*, Redmond: Tempus, 1987.

Pagels, Heinz R., *The Dreams of Reason: the computer and the rise of the sciences of complexity*, New York: Bantam, 1989.

Palfreman, Jon and Doron Swade, *The Dream Machine* London: BBC, 1991.

Pavel, Thomas G., *Fictional Worlds*, Cambridge. Mass.: Harvard University Press, 1986.

Peat, F. David, *Einstein's Moon*, Chicago: Contemporary Books, 1990.

Penrose, Roger, *The Emperor's New Mind: concerning computers, minds, and the laws of physics*, Oxford: Oxford University Press, 1989.

Poster, Mark, *The Mode of Information: poststructuralism and social context*, Cambridge: Polity Press, 1990.

Poundstone, William, *The Recursive Universe*, Oxford: Oxford University Press, 1987.

Prigogine, Ilya and Isabelle Stenger, *Order out of Chaos: man's new dialogue with nature*, New York: Bantam, 1984.

Rheingold, Harvey, *Virtual Reality*, London: Secker, 1991.

Rose, Frank, *West of Eden: the end of innocence at Apple Computer*, London: Hutchinson, 1989.

Roszak, Theodore, *The Cult of Information*, Cambridge: Lutterworth, 1986.

Rucker, Rudy, *Mind Tools*, London: Penguin, 1988.

Ryle, Gilbert, *The Concept of Mind*, London: Penguin, 1963.

Spence, Jonathan D., *The Memory Palace of Matteo Ricci*, London: Faber, 1985.

Stevens, Jay, *Storming Heaven: LSD and the American Dream*, London: Grafton, 1989.

Sutherland, Ivan, 'The ultimate display', *Proceedings of the International Federation of Information Processing Congress*, 1965, pp. 506–8.

Sutherland, Ivan, 'A head-mounted three-dimensional display', *Proceedings of the Joint Computer Conference*, 1968, vol. 33, pp. 757–64.

Turing, Alan, 'On computable numbers with an application to the entscheidungs problem', *Proceedings London Mathematical Society*, July 1937, vol. 42, pp. 230–65.

Turkle, Sherry, *The Second Self: computers and the human spirit*, New York: Simon & Schuster, 1985.

Virilio, Paul, *War and Cinema: the logistics of perception*, trans. Patrick Camiller, London: Verso, 1989.

Woodcock, Alexander and Monte Davis, *Catastrophe Theory*, London: Penguin, 1991.

Index

Discover more about our forthcoming books through Penguin's FREE newspaper...

Penguin
Quarterly

It's packed with:

- exciting features
- author interviews
- previews & reviews
- books from your favourite films & TV series
- exclusive competitions & much, much more...

Write off for your free copy today to:
Dept JC
Penguin Books Ltd
FREEPOST
West Drayton
Middlesex
UB7 0BR
NO STAMP REQUIRED

READ MORE IN PENGUIN

In every corner of the world, on every subject under the sun, Penguin represents quality and variety – the very best in publishing today.

For complete information about books available from Penguin – including Puffins, Penguin Classics and Arkana – and how to order them, write to us at the appropriate address below. Please note that for copyright reasons the selection of books varies from country to country.

In the United Kingdom: Please write to *Dept. JC, Penguin Books Ltd, FREEPOST, West Drayton, Middlesex UB7 OBR*

If you have any difficulty in obtaining a title, please send your order with the correct money, plus ten per cent for postage and packaging, to *PO Box No. 11, West Drayton, Middlesex UB7 OBR*

In the United States: Please write to *Penguin USA Inc., 375 Hudson Street, New York, NY 10014*

In Canada: Please write to *Penguin Books Canada Ltd, 10 Alcorn Avenue, Suite 300, Tororto, Ontario M4V 3B2*

In Australia: Please write to *Penguin Books Australia Ltd, 487 Maroondah Highway, Ringwood, Victoria 3134*

In New Zealand: Please write to *Penguin Books (NZ) Ltd, 182–190 Wairau Road, Private Bag, Takapuna, Auckland 9*

In India: Please write to *Penguin Books India Pvt Ltd, 706 Eros Apartments, 56 Nehru Place, New Delhi 110 019*

In the Netherlands: Please write to *Penguin Books Netherlands B.V., Keizersgracht 231 NL–1016 DV Amsterdam*

In Germany: Please write to *Penguin Books Deutschland GmbH, Friedrichstrasse 10–12, W–6000 Frankfurt/Main 1*

In Spain: Please write to *Penguin Books S. A., C. San Bernardo 117-6° E–28015 Madrid*

In Italy: Please write to *Penguin Italia s.r.l., Via Felice Casati 20, I–20124 Milano*

In France: Please write to *Penguin France S. A., 17 rue Lejeune, F–31000 Toulouse*

In Japan: Please write to *Penguin Books Japan, Ishikiribashi Building, 2–5–4, Suido, Tokyo 112*

In Greece: Please write to *Penguin Hellas Ltd, Dimocritou 3, GR–106 71 Athens*

In South Africa: Please write to *Longman Penguin Southern Africa (Pty) Ltd, Private Bag X08, Bertsham 2013*

READ MORE IN PENGUIN

SCIENCE AND MATHEMATICS

QED Richard Feynman
The Strange Theory of Light and Matter

'Physics Nobelist Feynman simply cannot help being original. In this quirky, fascinating book, he explains to laymen the quantum theory of light – a theory to which he made decisive contributions' – *New Yorker*

Does God Play Dice? Ian Stewart
The New Mathematics of Chaos

To cope with the truth of a chaotic world, pioneering mathematicians have developed chaos theory. *Does God Play Dice?* makes accessible the basic principles and many practical applications of one of the most extraordinary – and mind-bending – breakthroughs in recent years.

Bully for Brontosaurus Stephen Jay Gould

'He fossicks through history, here and there picking up a bone, an imprint, a fossil dropping and, from these, tries to reconstruct the past afresh in all its messy ambiguity. It's the droppings that provide the freshness: he's as likely to quote from Mark Twain or Joe DiMaggio as from Lamarck or Lavoisier' – *Guardian*

The Blind Watchmaker Richard Dawkins

'An enchantingly witty and persuasive neo-Darwinist attack on the anti-evolutionists, pleasurably intelligible to the scientifically illiterate' – Hermione Lee in the *Observer* Books of the Year

The Making of the Atomic Bomb Richard Rhodes

'Rhodes handles his rich trove of material with the skill of a master novelist ... his portraits of the leading figures are three-dimensional and penetrating ... the sheer momentum of the narrative is breathtaking ... a book to read and to read again' – Walter C. Patterson in the *Guardian*

Asimov's New Guide to Science Isaac Asimov

A classic work brought up to date – far and away the best one-volume survey of all the physical and biological sciences.

READ MORE IN PENGUIN

SCIENCE AND MATHEMATICS

The Panda's Thumb   Stephen Jay Gould

More reflections on natural history from the author of *Ever Since Darwin*. 'A quirky and provocative exploration of the nature of evolution ... wonderfully entertaining' – *Sunday Telegraph*

Einstein's Universe   Nigel Calder

'A valuable contribution to the demystification of relativity' – *Nature*. 'A must' – *Irish Times*. 'Consistently illuminating' – *Evening Standard*

Gödel, Escher, Bach: An Eternal Golden Braid
Douglas F. Hofstadter

'Every few decades an unknown author brings out a book of such depth, clarity, range, wit, beauty and originality that it is recognized at once as a major literary event' – Martin Gardner. 'Leaves you feeling you have had a first-class workout in the best mental gymnasium in town' – *New Statesman*

The Double Helix   James D. Watson

Watson's vivid and outspoken account of how he and Crick discovered the structure of DNA (and won themselves a Nobel Prize) – one of the greatest scientific achievements of the century.

The Quantum World   J. C. Polkinghorne

Quantum mechanics has revolutionized our views about the structure of the physical world – yet after more than fifty years it remains controversial. This 'delightful book' (*The Times Educational Supplement*) succeeds superbly in rendering an important and complex debate both clear and fascinating.

Mathematical Circus   Martin Gardner

A mind-bending collection of puzzles and paradoxes, games and diversions from the undisputed master of recreational mathematics.

READ MORE IN PENGUIN

SCIENCE AND MATHEMATICS

The Dying Universe Paul Davies

In this enthralling book the author of *God and the New Physics* tells how, from the instant of its fiery origin in a big bang, the universe has been running down. With clarity and panache Paul Davies introduces the reader to a mind-boggling array of cosmic exotica to help chart the cosmic apocalypse.

The Newtonian Casino Thomas A. Bass

'The story's appeal lies in its romantic obsessions ... Post-hippie computer freaks develop a system to beat the System, and take on Las Vegas to heroic and thrilling effect' – *The Times*

Wonderful Life Stephen Jay Gould

'He weaves together three extraordinary themes – one palaeontological, one human, one theoretical and historical – as he discusses the discovery of the Burgess Shale, with its amazing, wonderfully preserved fossils – a time-capsule of the early Cambrian seas' – *Mail on Sunday*

The New Scientist Guide to Chaos edited by Nina Hall

In this collection of incisive reports, acknowledged experts such as Ian Stewart, Robert May and Benoit Mandelbrot draw on the latest research to explain the roots of chaos in modern mathematics and physics.

Innumeracy John Allen Paulos

'An engaging compilation of anecdotes and observations about those circumstances in which a very simple piece of mathematical insight can save an awful lot of futility' – Ian Stewart in *The Times Educational Supplement*

Fractals Hans Lauwerier

The extraordinary visual beauty of fractal images and their applications in chaos theory have made these endlessly repeating geometric figures widely familiar. This invaluable new book makes clear the basic mathematics of fractals; it will also teach people with computers how to make fractals themselves.